Climate Change Mitigation, Technological Innovation and Adaptation

THE FONDAZIONE ENI ENRICO MATTEI (FEEM) SERIES ON ECONOMICS, THE ENVIRONMENT AND SUSTAINABLE DEVELOPMENT

FEEM is a nonprofit, nonpartisan research institution devoted to the study of sustainable development and global governance. Founded by the Eni group, officially recognized by the President of the Italian Republic in 1989, and in full operation since 1990, FEEM has grown to become a leading research centre, providing timely and objective analysis on a wide range of environmental, energy and global economic issues.

FEEM's mission is to improve – through the rigor of its research – the credibility and quality of decision-making in public and private spheres. This goal is achieved by creating an international and multidisciplinary network of researchers working on several innovative projects, by providing and promoting training in specialized areas of research, by disseminating research results through a wide range of outreach activities, and by delivering directly to policy makers via participation in various institutional fora.

The Fondazione Eni Enrico Mattei (FEEM) Series on Economics, the Environment and Sustainable Development publishes leading-edge research findings providing an authoritative and up-to-date source of information in all aspects of sustainable development. FEEM research outputs are the results of a sound and acknowledged co-operation between its internal staff and a worldwide network of outstanding researchers and practitioners. A Scientific Advisory Board of distinguished academics ensures the quality of the publications.

This series serves as an outlet for the main results of FEEM's research programmes in the areas of economics, the environment and sustainable development.

Titles in the series include:

The Social Cost of Electricity
Scenarios and Policy Implications
Edited by Anil Markandya, Andrea Bigano and Roberto Porchia

Climate Change Mitigation, Technological Innovation and Adaptation
A New Perspective on Climate Policy
Edited by Valentina Bosetti, Carlo Carraro, Emanuele Massetti and Massimo Tavoni

Climate Change Mitigation, Technological Innovation and Adaptation

A New Perspective on Climate Policy

Edited by

Valentina Bosetti
Bocconi University, Fondazione Eni Enrico Mattei and Euro-Mediterranean Centre on Climate Change, Italy

Carlo Carraro
Ca' Foscari University of Venice, Fondazione Eni Enrico Mattei and Euro-Mediterranean Centre on Climate Change, Italy

Emanuele Massetti
Fondazione Eni Enrico Mattei and Euro-Mediterranean Centre on Climate Change, Italy

Massimo Tavoni
Fondazione Eni Enrico Mattei and Euro-Mediterranean Centre on Climate Change, Italy

THE FONDAZIONE ENI ENRICO MATTEI (FEEM) SERIES ON ECONOMICS, THE ENVIRONMENT AND SUSTAINABLE DEVELOPMENT

Edward Elgar
Cheltenham, UK • Northampton, MA, USA

Published by
Edward Elgar Publishing Limited
The Lypiatts
15 Lansdown Road
Cheltenham
Glos GL50 2JA
UK

Edward Elgar Publishing, Inc.
William Pratt House
9 Dewey Court
Northampton
Massachusetts 01060
USA

A catalogue record for this book
is available from the British Library

Library of Congress Control Number: 2014932592

This book is available electronically in the ElgarOnline.com Social and Political Science Subject Collection, E-ISBN 978 1 78347 717 3

ISBN 978 1 84980 949 8

Printed and bound in Great Britain by T.J. International Ltd, Padstow

Contents

Contributors

Valentina Bosetti, Bocconi University, Fondazione Eni Enrico Mattei and Euro-Mediterranean Centre on Climate Change, Italy.

Carlo Carraro, Ca' Foscari University of Venice, Fondazione Eni Enrico Mattei and Euro-Mediterranean Centre on Climate Change, Italy.

Enrica De Cian, Fondazione Eni Enrico Mattei and Euro-Mediterranean Centre on Climate Change, Italy.

Thomas Longden, Fondazione Eni Enrico Mattei and Euro-Mediterranean Centre on Climate Change, Italy.

Emanuele Massetti, Fondazione Eni Enrico Mattei and Euro-Mediterranean Centre on Climate Change, Italy.

Lea Nicita, Fondazione Eni Enrico Mattei, Italy.

Fabio Sferra, Fondazione Eni Enrico Mattei, Italy.

Alessandra Sgobbi, Fondazione Eni Enrico Mattei, Italy, currently at European Commission, Joint Research Centre, Institute for Energy and Transport.

Massimo Tavoni, Fondazione Eni Enrico Mattei and Euro-Mediterranean Centre on Climate Change, Italy.

Introduction

Valentina Bosetti, Carlo Carraro, Emanuele Massetti and Massimo Tavoni

Although climate change is recognised by the scientific community as a major threat to ecosystems and human wellbeing, little progress has been made in regulating its source, and actually emissions of greenhouse gases (GHG) are continuously on the rise (Boden et al., 2011). The global economic crisis has only temporarily slowed the growth in emissions, and global emissions have increased by more than 5 per cent in 2010, an unmatched record in the last two decades (Friedlingstein et al., 2010; Hansen et al., 2011). Concentrations of carbon dioxide in the atmosphere have reached 397 parts per million (ppm) in July 2013.[1] They are increasing rather steadily at about two parts per million (ppm) per year. Since carbon dioxide remains in the atmosphere for centuries, each additional tonne of carbon dioxide emitted now will affect the wellbeing of people that will inhabit our planet decades and centuries from now. Concentrations in the atmosphere of other GHG gases are also increasing steadily, exacerbating the problem.

As a result, the atmosphere traps more heat and the global average surface temperature is increasing. Recent estimates indicate that the average surface temperature has increased by about 0.6 degrees Celsius (°C) with respect to 1951–1980, about 0.8°C with respect to the pre-industrial average. Temperatures will continue to rise for decades because the climate system has a delayed response to the stock of GHG, and equilibrium temperature grows linearly with cumulative emissions of CO_2. It is very likely that we will live in a warmer planet for several centuries. The effects of this warming will be spread unevenly across the globe, depending on the variation in regional and local climatic effects and on the differences in vulnerability that characterises societies.

The long-term nature of the problem, the need for coordinated action, as well as the uneven and uncertain distribution of costs and benefits of climate policies explain the deadlock of international climate negotiations. Even assuming an implausible immediate and universal agreement to curb GHG emissions, warming will occur, and likely cross the 2°C signpost.

It is possible that technological breakthroughs will make the transition to a zero-emissions society easier than what we anticipate today. It could even be that absorbing emissions to reduce the stock of GHG in the atmosphere on a large scale might become possible at reasonable cost. Although there is room for optimism, in order to produce robust policy making recommendations it is crucial to carefully analyse environmental and economic impacts of a large set of scenarios, spanning different climate policies, degrees of international commitments, technological scenarios, actions (i.e. innovation, mitigation and adaptation), policy mechanisms and enforcement measures. Back of the envelope calculations and expert judgment can provide useful guidance to inform potential strategies to control climate change, but the complexity of the problem and the high stakes involved commands an integrated approach which can quantify the trade-offs faced by policy.

In this book we present a state of the art integrated assessment model, WITCH, which has been developed with this complex type of analyses in mind, and a broad set of scenarios that have been explored using this tool. Most of the material that is collected in this book has already been published in international peer-reviewed journals, as chapters in books, or as working papers. However, we believe it is valuable to collect and synthesise these studies and fragments of research in a single narrative.

We wrote the first equations of WITCH in 2005 and since then the model has grown in complexity and richness. However, the core of the model has not changed because we built it using the following five necessary ingredients for a sound analysis of climate change policy:

- A long-term economic growth framework;
- A global scope with regional detail;
- A description of the interaction among world regions that can capture free riding incentives;
- A compact and tractable representation of the main climate mitigation options;
- An endogenous representation of technological change.

These features are reflected in the name itself: World Induced Technical Change Hybrid model, or WITCH.

WITCH is a hybrid integrated assessment model. Integrated assessment models describe the full cycle of environmental externalities. In WITCH, economic activity generates greenhouse gas emissions, a simplified model of carbon and climate dynamics translates emissions in to temperature increases and a damage functions provides a feedback of climatic change on economic activities (Figure 1).

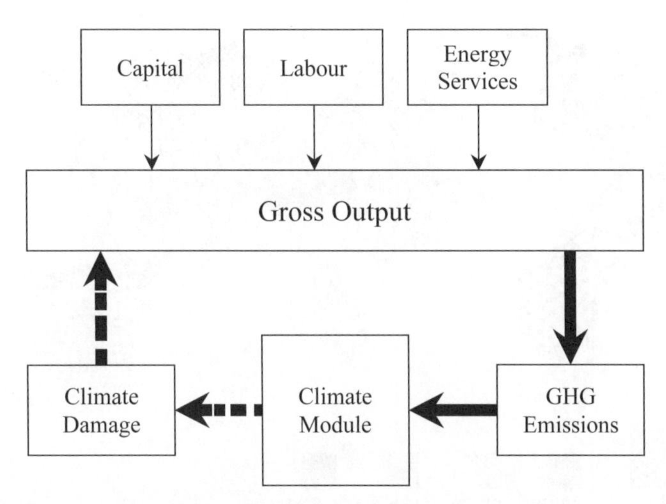

Figure 1 WITCH as an integrated assessment model

The model is defined as hybrid as it merges a top-down model of economic growth, describing each region, with a bottom-up compact representation of the mitigation strategies. These two cores of the economy are completely integrated. The allocation of investments across technologies and uses is defined as an optimal intertemporal strategy.

As climate change is a global externality, we adopt this hybrid set up to model the economic systems of the whole world, divided into 12 macro regions as shown in Figure 2 (the latest version of the code has 14 regions).

Each region (or coalition of regions) maximises its own welfare, given a set of production constraints, information about technologies, the budget constraint and the behaviour of other regions, which in turn affect global variables, such as global temperature. Therefore, the investment profile for each technology in each region is the solution of an intertemporal game between the regions (or coalitions of regions). More generally, these regions behave strategically with respect to all decision variables, playing a game with an open-loop information structure. Therefore the model captures the two dimensions along which agents behave strategically: the geographical dimension (rich vs. poor, high damage vs. low damage regions, etc.) and the time dimension (present vs. future generations).

More specifically WITCH can be viewed as a game: the players are the regions, the actions that each player has is the path of investments, the outcome is a consumption path over the whole simulation horizon, and the rules prescribe a non-cooperative, simultaneous, open membership, game with non-orthogonal free-riding. WITCH can simulate all degrees of cooperation among the macro-regions in which world countries are

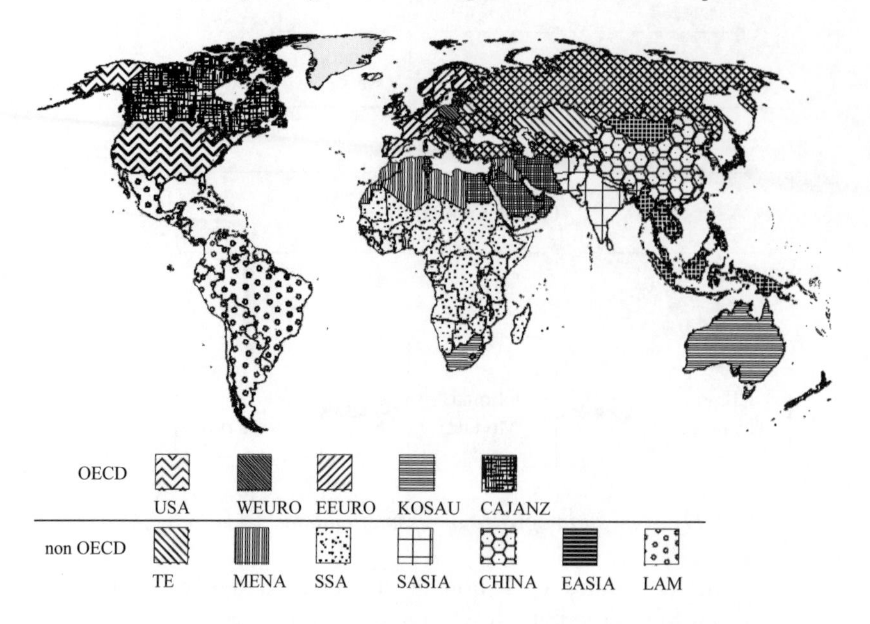

Notes: USA: United States of America; WEURO: Western EU countries; EEURO: Eastern EU countries; KOSAU: South Korea, South Africa, Australia; CAJANZ: Canada, Japan, New Zealand; TE: Non-EU Eastern European countries, including Russia; MENA: Middle East and North Africa; SSA: Sub Saharan Africa; SASIA: South Asia including India; CHINA: China including Taiwan; EASIA: South East Asia; LAM: Latin America, Mexico and Caribbean.

Figure 2 WITCH the regional dimension

aggregated. At one extreme we find a world in which regions do not cooperate and act only in self interest. This is the decentralised or non-cooperative solution. At the other extreme there is a world in which countries can form coalitions to control optimally global externalities, one or more at a time. In between these two extremes we find all possible combinations of smaller coalitions that coexist with non-cooperating regions, as well as cases in which only a subset of externalities are internalised.

The evolution of technology plays a crucial role when analysing long term growth and the potential for mitigation and adaptation. WITCH tries to capture the essence of this by mimicking both innovation and diffusion processes. As knowledge is, at least partially, a non-rival good, spillovers across regions play a crucial role. Indeed, they introduce a second market failure, in addition to the standard climate externality.

Thus, WITCH's unique feature is that it can represent multiple externalities, with consequent important implications for economic policy. The model can be used to show the implications of unregulated markets in the presence of externalities, and study the effectiveness of policy tools aimed at improving welfare, both from a regional and a global perspective. For this reason the WITCH model was used to evaluate climate policies in several contexts, from research projects, to consultancies and briefings to governmental as well as non-governmental organisations, and as an active member of the international integrated assessment modelling community.

This book is intended for the reader who wants to explore the world of integrated assessment modelling, understand its policy relevance and research potential, either as an external user of model results or directly as a modeller and a researcher. Some background on climate change economics is expected, as well as some rudimental knowledge of integrated assessment modelling, but most of the chapters in this book are relatively easy to follow and to interpret.

The rest of the book is structured into seven chapters, each of them centred on a broad area of research. In a nutshell, Chapters 2–5 mainly deal with issues related to climate change mitigation. Chapter 6 opens up the issue of adaptation to climate change and its impacts. Chapter 7 lays out the future ahead and the main ongoing research focuses.

More in detail, Chapter 2 sets up the stage by framing the broader picture of climate change mitigation, both in terms of the technological implications and the economic cost of reducing Greenhouse Gas (GHG) emissions. It shows how constraints on key technologies, the time and international scope of climate policy affect the optimal mix of energy investments in the energy sector, in energy R&D, and the overall cost of the mitigation policy. It is important to stress that an optimal solution is one that maximises the welfare of regions, or of coalitions of regions, given all the constraints imposed in the model.[2] It is not a first-best. Chapter 2 also shows how ambitious mitigation targets might not be feasible if countries delay emission reductions or only a fraction of countries reduce emissions.

Chapter 3 provides a focus on technological innovation. It illustrates several developments to the model that have looked in a wider sense to the role of directed technical change and endogenous growth. The chapter summarises how climate policy should be designed to incorporate main policy lessons from these exercises.

Chapter 4 explains why international cooperation on climate policy is weak. WITCH was used to replicate different degrees of cooperation among countries, from nationalistic free-riding behaviour, to partial and global cooperation in order to assess incentives faced by different regions of the world. Equity and distributional issues are at the core of the work presented in Chapter 4.

Chapter 5 enlarges the picture to include uncertainty. Uncertainty surrounding climate change, its effect but also its solutions, should have a major role in designing effective and adaptive climate policies. WITCH was used to examine optimal strategies to hedge against these sources of uncertainty and this chapter summarises the main lessons that can be drawn from these exercises.

Forests are a major sink of CO_2 and their management has been recognised has having a paramount importance within any international climate policy. Chapter 6 reports the work done so far coupling the WITCH model with state of the art forestry model in order to assess the potential benefits that may derive from protecting and enhancing global forest carbon stocks. The chapter also discusses the other side of the coin, that is, by lowering the pressure on the energy sector, the inclusion of forest emissions into a climate policy changes the incentive to invest in carbon-free technologies and in innovation.

As climate policy cannot only deal with mitigation, but has to look into adaptation costs and potentials as well, Chapter 7 reports on the developments to the WITCH model that aim at incorporating adaptation and assess the trade off with mitigation initiatives. The basic idea is that the welfare maximising level of emissions should be determined by simultaneously solving the mitigation and the adaptation problem. The objective is to control climate change impacts rather than global temperature or climate change.

Chapter 8 illustrates the road ahead by discussing the most recent developments to the model, the frontier of integrated assessment analysis and where this might bring us.

WITCH is an open project and new additions as well as developments will follow in the next years. A final consolidated version will probably never exist. The latest developments will be posted at the WITCH project website (www.witch-model.org). Researchers and policy makers will be able to consult all the background material and also the output of a number of representative scenarios using the WITCH Policy Simulator (www.witch-model.org/simulator), a section of the website that provides user-friendly access to an interesting dataset. The model code is also available for download, as a way to foster community research in the field.

We hope that researchers and policy makers who are interested in applied climate economics will find this book a source of useful information. We are perfectly aware that numerical tools like WITCH have many limits and must be applied with caution. However, they are extremely valuable to appreciate the many subtleties of the climate change challenge. Interdisciplinary and applied economic work in this field is exciting and to a great extent new: much more work lies ahead of us all in order to carry out analyses which is informative and insightful.

ACKNOWLEDGEMENTS

The development of the model would have not been possible without the generous support of funding and grants from FEEM, CMCC, the European Commission (PASHMINA, TRANSUST, TRANSUST-SCAN) and the OECD. We are deeply grateful to Marzio Galeotti for his contribution to the development of the early stages of the WITCH model. We are also indebted to Alessandro Lanza for strong support to the project, for many insights and helpful comments.

NOTES

1. Data from the observatory of Mauna Loa, http://www.esrl.noaa.gov/gmd/ccgg/trends/ consulted on September 4th, 2013.
2. In the book optimal, best, cost-minimising are used interchangeably.

REFERENCES

Boden, T.A., G. Marland and R.J. Andres (2011), 'Global, regional, and national fossil-fuel CO_2 emissions', Carbon Dioxide Information Analysis Center, Oak Ridge National Laboratory, US Department of Energy, Oak Ridge, Tenn., USA, doi 10.3334/CDIAC/00001_V2011.

Friedlingstein P., R.A. Houghton, G. Marland, J. Hacker, T.A. Boden, T.J. Conway, J.G. Canadell, M.R. Raupach, P. Ciais and C. Le Quéré (2010), 'Update on CO_2 emissions', *Nature Geoscience*, **3**, 811–812, doi 10-1038/ngeo1022.

Hansen, J.E., R. Ruedy, M. Sato and K. Lo (2011), 'NASA GISS surface temperature (GISTEMP) analysis', in *Trends: A Compendium of Data on Global Change*, Carbon Dioxide Information Analysis Center, Oak Ridge National Laboratory, US Department of Energy, Oak Ridge, Tenn., USA. doi: 10.3334/CDIAC/cli.001.

1. A Climate-Constrained World

Emanuele Massetti

Climate change may harm future generations. According to the latest IPCC report (IPCC, 2007), anthropogenic emissions of greenhouse gases (GHG) are among the main causes of climate change, even though uncertainty remains as to their exact relevance in the overall climatic process: thus it is necessary to identify how, when, and where these emissions ought to be controlled in order to avoid dangerous climate changes.[1]

At the 2008 G8 Summit in Japan, the leading industrialised nations agreed on the objective of at least halving global CO_2 emissions by 2050. G8 and Major Economies Forum (MEF)[2] leaders reiterated this target in Italy, in July 2010, specifying that richer economies should commit to at least an 80 percent reduction in 2050.

These long-term declarations of intents clash with the difficulties the international community has in establishing a global agreement with binding short- and mid-term emission reduction targets. The Conference of the Parties (COP) held in Copenhagen in December 2009 failed to produce a global mandatory regime to cut GHG emissions. A more modest Copenhagen agreement collected non-binding pledges to reduce GHG emissions in 2020, which appear to have a modest effect on future emissions levels (UNEP 2010, Höhne et al., 2012). More relevant was the commitment to devote a substantial amount of resources to finance adaptation and mitigation in developing countries (Carraro and Massetti 2011a). The Copenhagen Pledges and the financial provisions of the Copenhagen accord were incorporated into the UNFCCC legal framework in December 2010, by the COP held in Cancun, Mexico. The COP in Durban in December 2011 made progress in defining important issues related to climate finance, but the parties postponed any decision for the post 2020 climate regime to an ad-hoc group called the Durban Platform, that is in charge of developing a successor to the Kyoto Protocol by 2015. In Doha, in 2012, the parties defined the rules governing the second commitment period for the Kyoto Protocol. Only the European Union and Australia have emission reduction targets from 2013 to 2020 while Canada, Japan and Russia abandoned the treaty.

The fragmented international climate architecture leaves space to single regions, countries, and local actors to adopt their own emission reduction targets. For example, the European Union already has a mandatory goal to reduce emissions by 20 percent in 2020, with respect to 1990.[3] In November 2011 the Australian Parliament introduced a tax of AUD23 on about 60 percent of carbon emissions. The tax will increase by 2.5 percent annually until 2015, when it will evolve into a cap-and-trade system with a price corridor. In the USA the chances of soon having a national legally binding emissions target are extremely low, but several States are taking steps to promote emissions reductions and energy savings. For example, California's Global Warming Solutions Act of 2006 limits California's GHG emissions in 2020 to their 1990 level.

Consequently, there is increasing interest in, and a need for, research efforts providing information on the best strategy that different regions of the world should adopt in order to minimise the cost of achieving their own emission reduction target. In particular, it is crucial to identify the long-term investment mix in the energy sector in different world regions, taking into account the role of technical progress and the future evolution of different technologies.

This chapter concisely exposes lessons learnt from a large body of research on optimal mitigation strategies generated using the WITCH integrated assessment model. In particular, this chapter reviews a set of scenarios developed using the same version of the WITCH model. The Business-as-Usual scenario is compared to 11 climate policy scenarios in which global GHG concentrations are constrained to remain below 535 and 640 ppm CO_2-eq in 2100. Throughout this and the following chapters, stabilisation targets are expressed as concentration of Kyoto gases, unless otherwise specified.

The 535 stabilisation target is examined under a series of technological and policy constraints. The analysis covers the implications of limiting the deployment of key low-emission technologies, for instance coal with carbon capture and sequestration (CCS), nuclear, renewables. It also illustrates how the optimal energy mix and mitigation costs change when countries delay actions to reduce emissions or when some countries do not make any effort at all to contain global warming. The scenarios can be downloaded from the website www.witchmodel.org. The website hosts the WITCH Policy Simulator, a user-friendly interface to visually explore, compare, and download data from the scenarios.

Despite stabilising GHG concentrations at 535 ppm CO_2-eq in 2100 requires remarkable transformations, temperature is expected to increase well above the 2°C target indicated in the Cancun Agreement. Since concentrations of GHG in the atmosphere are already very close to 450 ppm CO_2-eq, it is possible to achieve the target only using technologies that absorb GHG from the atmosphere. The most attractive options are electricity generation with biomass and CCS and direct air capture, not included in the version of the model used in this chapter. The latest version of the model features both these technologies and it has already been used to generate scenarios of stabilisation at 450 ppm CO_2-eq (Massetti, 2011; Massetti and Tavoni, 2012; Carraro et al., 2012; Bosetti et al., 2012). Chapter 7 illustrates these new model developments.

These scenarios have constituted the bulk of the modeling effort in the past two years, since all integrated assessment models – including WITCH- have focused on 2°C compatible policies. However, in the analysis presented in this chapter no such stringent scenarios is included. Given the very stringent nature of 450 ppm-eq scenarios, they are best dealt with in model intercomparison projects, in which several different integrated assessment models run the same set of scenarios, thus allowing to identify results which are robust to model specification. We refer the interested reader to the forthcoming 5th assessment report of the IPCC, a large fraction of which will be devoted to the assessment of 2°C policies, with an important scenario contribution of WITCH.

The objective of this chapter is to illustrate how a constraint on emissions changes the optimal allocation of investments in energy technologies, the energy and carbon intensity of the economies, the amount of resources to invest in Research and Development (R&D) and long-term economic growth.

The WITCH model yields the equilibrium intertemporal allocation of investments in energy technologies and R&D that belong to the best economic and technological responses to different policy measures. The game theory set-up accounts for interdependencies and spillovers among 12 regions of the world. Therefore, equilibrium strategies are the outcome of a dynamic game through which inefficiencies induced by global strategic interactions can be assessed. In WITCH, technological progress in the energy sector is endogenous, thus accounting for the effects of different stabilisation scenarios on induced technical change, via both innovation and diffusion processes. Feedback from economic to climatic variables, and vice versa, is also accounted for in the dynamic system.

These features enable WITCH to address many questions that naturally arise when analysing carbon mitigation policies. Among these questions, this chapter aims to answer the following:

- What are the implications of the proposed stabilisation targets for investment strategies and consumption of traditional energy sources *vis à vis* low carbon options?
- What is the optimal balance between mitigation measures in the energy sector, to reduce emissions from Land Use, Land Use Change

and Forestry (LULUCF) and non-CO_2 GHG emissions?
- What is the role of public energy R&D expenditures for generating improvements in both energy efficiency and carbon intensity?
- How sensitive are the economic costs of climate policies to different technological scenarios, and in particular, to hypotheses on major technological breakthroughs?

The structure of the chapter is as follows. Section 1.1 describes the scenarios, the policy framework, and the stabilisation targets. Section 1.2 illustrates the optimal path of investments to reduce emissions in the energy sector. Section 1.3 examines GHG abatement opportunities in the non-energy sector. Section 1.4 focuses on innovation strategies. Section 1.5 provides estimates of the economic costs of climate policy with a focus on technological choices. A final section concludes the chapter.

1.1 SCENARIOS OF CLIMATE POLICY: TARGETS AND CONSTRAINTS

This section illustrates the best response strategies to achieve two stabilisation targets. The first target requires that GHG concentrations in the atmosphere do not exceed 640 ppm CO_2-eq. The second target is more stringent and requires concentrations to remain below 535 ppm CO_2-eq. With the first target global mean surface temperature increases 3.7°C above the pre-industrial level, in 2100. With the second target the temperature increases by 2.5°C in 2100. Since the climate system has a strong inertia, temperature continues to rise in the 640 ppm CO_2-eq scenario after 2100 while it reaches a plateau in the 535 ppm CO_2-eq scenario. Nevertheless, the 535 ppm CO_2-eq scenario is not sufficient to achieve the 2°C temperature target. As all other IAMs, WITCH can achieve the 2°C target only if it is possible to absorb large quantities of GHG from the atmosphere (see for example Clarke et al., 2009). Chapter 7 illustrates recent work in this area. This section focuses on the 535 ppm CO_2-eq target because it is more policy relevant than the 640 target and illustrates how alternative assumptions on climate policy, on international cooperation, and on the availability of key technologies affect the optimal mix of investments and the cost of achieving the target (see Table 1.1). Figure 1.1 displays a first interesting insight: if countries delay the reduction of GHG emissions to 2020, the temperature will be slightly higher than 2.5°C during the transition to the equilibrium. Further delays, even if compensated by more aggressive action in the future, might lead to overshooting the long-term temperature target.

Figure 1.2 shows Business-as-Usual (BaU) emissions together with

Table 1.1 Description of scenarios

Scenario name	Stabilization target in 2100 (ppm CO_2-eq)	Peak of emissions (year)	Global participation	Technology with constraints	Constraints on emission trading
BaU	None	2100	·	None	·
640	640	2030	Yes	None	No
535	535	2010	Yes	None	No
535 – Limited Wind & Solar	535	2010	Yes	Wind & Solar	No
535 – Limited R&D	535	2010	Yes	R&D for energy efficienty	No
535 – Limted nuclear	535	2010	Yes	Nuclear	No
535 – Limited CCS	535	2010	Yes	CCS	No
535 – No backstop	535	2010	Yes	Backstop	No
535 – No redd	535	2010	Yes	Redd	No
535 – Start in 2020	535	2020	Yes	None	No
535 – No China & India	535	2010	China and India join in 2030	None	No
535 – Limited trade	535	2010	Yes	None	Yes

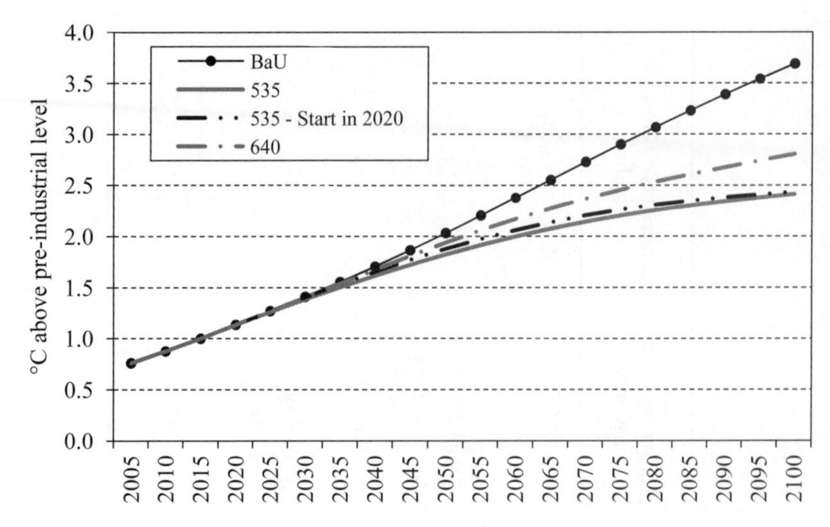

Figure 1.1 Global mean temperature

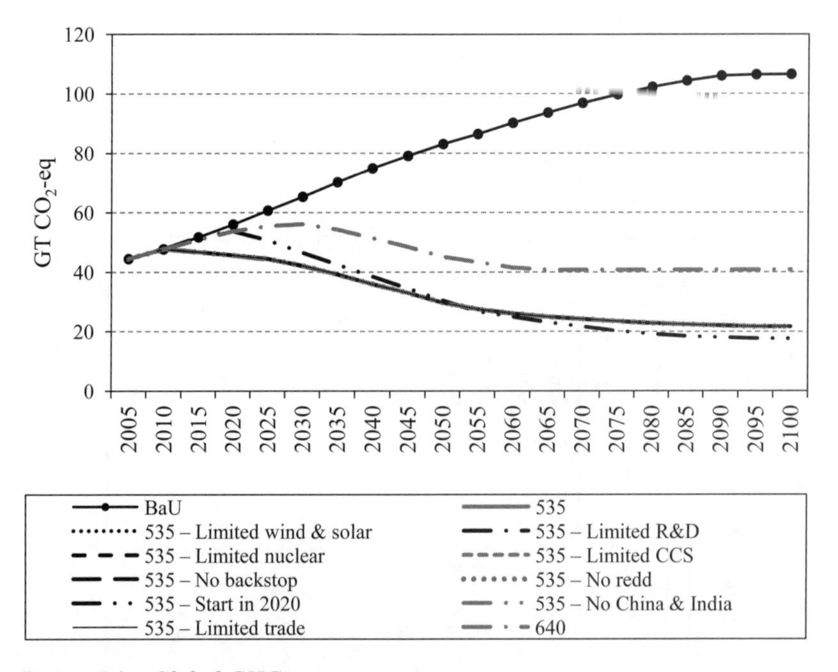

Figure 1.2 Global GHG emissions

emission time profiles for the two stabilisation targets. These are optimal time profiles because they were obtained by computing the fully cooperative equilibrium of the game given the GHG concentration constraints, by solving a global joint welfare maximisation problem where environmental externalities are internalised. Note that feedbacks from climate damage to the production of economic goods are taken into account when computing the optimal emission profiles.

Without any stabilisation policy (the BaU scenario), GHG emissions reach 106 Gt CO_2-eq/yr in 2100. In the 535 ppm CO_2-eq scenario, annual emissions peak in 2010 and reach 22 CO_2-eq in 2100. If the target is 640 ppm CO_2-eq, GHG emissions peak at 56 Gt CO_2-eq/yr in 2030 and then start to decline until they reach 41 CO_2-eq in 2100. In the 535 ppm target with delayed action, emissions peak in 2020, rather than in 2010, and then decline at faster pace and must be equal to 18 Gt CO_2-eq/yr in 2100. The emission reductions required to meet the more stringent stabilisation target are particularly challenging, given the expected growth rate of world population and GDP: per capita emissions in 2100 would have to decline from about 11.7 tCO_2-eq/cap per year to 2.4 tCO_2-eq/cap per year. If actions are delayed by only 10 years, emissions per capita in 2100 must be equal to 1.9 tCO_2-eq/cap per year, 25 percent lower than in the case with immediate action.

To achieve the two stabilisation targets and the related optimal emission profile, it is assumed that all regions of the world agree on implementing a cap-and-trade policy. This is certainly a strong simplification, but it allows to endogenously derive global marginal abatement cost curves and to study the implications for investment portfolios. Banking and borrowing are not allowed.[4] The model is solved as a game. Each region of the world maximises its own welfare given the strategies adopted in the other regions of the world. An iterative process leads to the unique intertemporal Nash equilibrium of the game, which defines the equilibrium investment strategies in each world region, in all periods of time.

1.2 THE ENERGY SECTOR

All the scenarios reveal that climate policy induces two major changes in the energy sector. There is indeed the incentive to switch from fuels with a high carbon content (coal) to fuels with low carbon content (natural gas, biomass) and to adopt energy technologies that reduce emissions (carbon capture and storage – CCS) or have zero/limited emissions (renewable, nuclear). All these actions reduce the content of carbon per unit of energy, i.e. the carbon intensity of energy. However reducing the carbon intensity of energy is costly. Therefore climate policy creates the incentive to save energy use per unit of output, to reduce the energy intensity of the economy. All the scenarios show that it is optimal to strongly reduce both the carbon intensity of energy and the energy intensity of GDP. However, while the dynamic of factor prices and of technical progress in the BaU scenario induces strong energy efficiency improvements, the lack of any constraint on GHG emissions, the abundance of fossil fuels (especially coal), and the growth of coal-hungry Asian economies induce an increase of the global carbon intensity of energy. Therefore all the scenarios unanimously convey the key message that first and foremost, climate policy requires countries to depart from the use of fossil fuels. However, the second key message that emerges from the analysis is that actions to reduce the energy intensity of the economy are less expensive and should be implemented before investing in costly measures to decarbonise energy. Table 1.2 also reveals that a moderate stabilisation target such as the 640 ppm CO_2-eq, still requires a massive change in the energy sector, capable of halving the carbon content of energy compared to the current level in 2100.

For the same stabilisation target the lack of carbon-free electricity generation technologies requires a higher effort to increase the energy efficiency. Postponing mitigation action requires faster and deeper cuts in both energy intensity and carbon intensity. Incomplete participation determines a greater incentive to reduce the energy intensity in the first years for the participating countries.

The dynamic paths of energy intensity and carbon intensity of energy implied by the two stabilisation scenarios require drastic changes in the energy sector. The rest of this section analyses the investment paths in different energy technologies.

The energy sector is characterised by long-lived capital. Therefore the investment strategies pursued in the next two to three decades will be crucial in determining the emissions pathways that will eventually emerge in the second half of the century. All the scenarios show that it is necessary to develop a new strategy in the energy sector, targeted mainly to decarbonise energy production. This can be done through the extensive deployment of currently known abatement technologies and or through the development of new energy technologies.

Supply cost curves of abatement vary widely across sectors; for example they are believed to be especially steep in the transport sector. Power generation is comparatively more promising: it is a heavy weight sector in terms of emissions and one of the few for which alternative production technologies are available. Not surprisingly, all the scenarios show a significant contribution of electricity in mitigation, as illustrated in Figure 1.3. To optimally achieve a 535 ppm CO_2-eq concentration target, almost all electricity (around 80 percent) will have to be generated at low, almost zero, carbon rates by 2050. The milder 640 ppm CO_2-eq target allows a more gradual transition away from fossil fuel based electricity, but nonetheless shows a

Table 1.2 Energy intensity of GDP and carbon intensity of energy

	BaU	535	535 – Limited wind & solar	535 – Limited R&D	535 – Limited nuclear	535 – Limited CCS	535 – No backstop	535 – No REDD	535 – Start in 2020	535 – No China & India	535 – Limited trade	640
Index of energy intensity (base 2005)												
2025	0.74	0.64	0.63	0.63	0.62	0.63	0.56	0.54	0.61	0.53	0.57	0.71
2050	0.51	0.36	0.35	0.36	0.33	0.35	0.24	0.27	0.29	0.28	0.28	0.40
2075	0.39	0.29	0.25	0.29	0.27	0.31	0.19	0.20	0.19	0.21	0.19	0.30
2100	0.32	0.29	0.22	0.28	0.28	0.31	0.18	0.18	0.19	0.18	0.16	0.27
Index of carbon intensity (base 2005)												
2025	1.04	0.91	0.92	0.91	0.91	0.92	0.92	0.89	0.98	0.91	0.91	1.01
2050	1.11	0.51	0.52	0.51	0.51	0.52	0.57	0.48	0.53	0.51	0.51	0.79
2075	1.16	0.31	0.36	0.32	0.32	0.29	0.34	0.28	0.26	0.32	0.32	0.61
2100	1.16	0.22	0.28	0.22	0.22	0.21	0.23	0.20	0.15	0.22	0.22	0.52

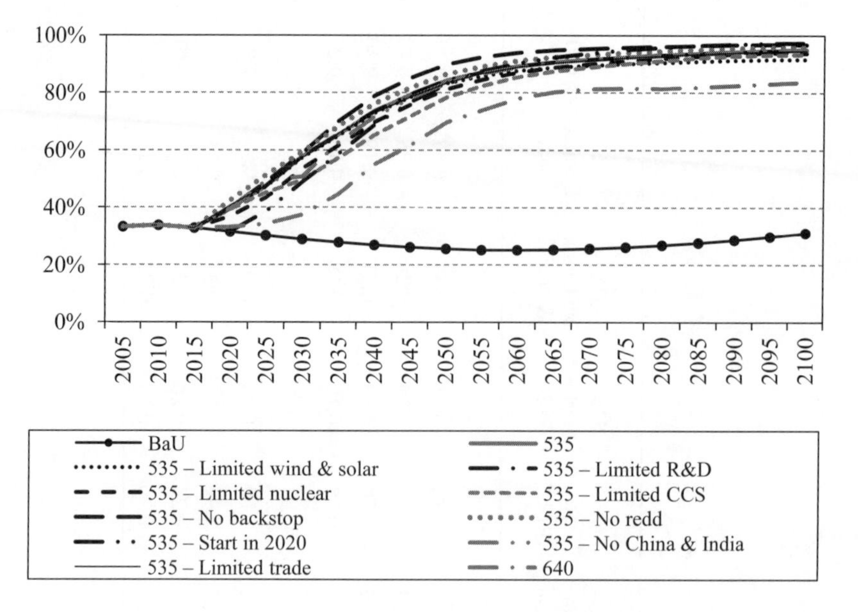

Figure 1.3 Share of zero or low carbon electricity generation

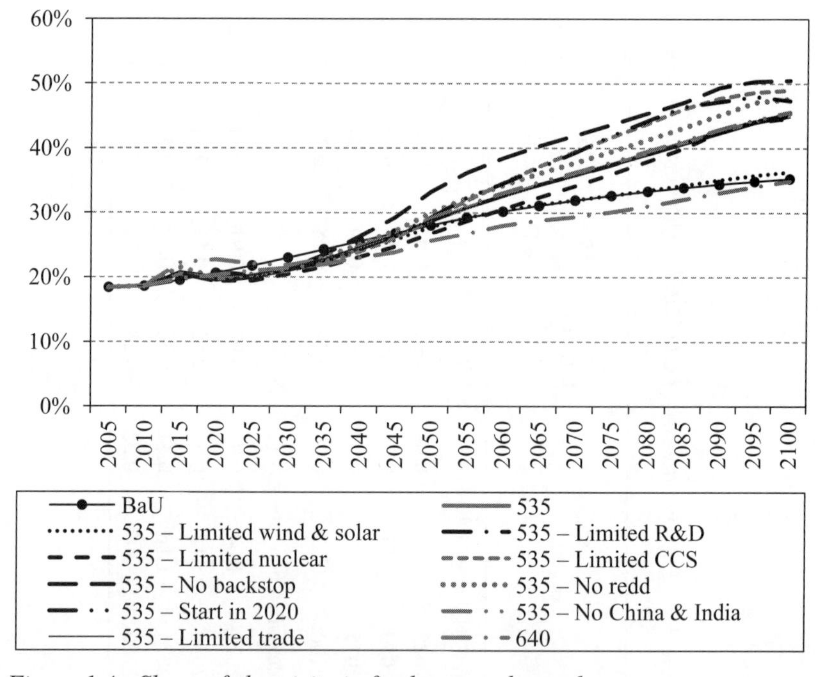

Figure 1.4 Share of electricity in final energy demand

noticeable departure from the BaU scenario. The model shows that virtually all electricity must come from low or zero emissions sources in 2100.[5]

The scenario in which there are limits to the penetration of backstop energy experiences a higher penetration of low or zero carbon emissions electricity because the absence of a carbon-free alternative in the transport sector (backstop non-electric) increases the incentive to decarbonise the electricity sector. In scenarios without the possibility to reduce emissions from deforestation and land degradation (No REDD), with delayed action (Start in 2020) and with partial cooperation (No China & India) it is also optimal to have higher than average penetration of zero or low emission power generation technologies.

The possibility to drastically reduce the emissions in the power sector increases the attractiveness of electricity compared to other final energy sources (Figure 1.4). The share of final energy demand detained by electricity increases over time in the BaU scenario and it is optimal to increase it further under all 535 ppm CO_2-eq stabilisation targets, except for the scenario with limits to the penetration of renewables (Limited wind & solar).

Figure 1.5 illustrates the electricity mix in the BaU scenario and in all stabilisation scenarios. Climate policy transforms the optimal electricity mix: fossil fuels are quickly replaced by renewables and other low-emissions power generation technologies. In all 450 ppm CO_2-eq scenarios, coal without CCS is rapidly phased-out. Coal with CCS plays a relevant role at global level but the high cost of the power plants, the gradual exhaustion of the less expensive reservoirs and the leakage of carbon during the capture process (10 percent) limit its expansion. Coal with CCS has greater relevance in the 640 ppm CO_2-eq scenario than in the 535 ppm CO_2-eq scenario because leakage of carbon is less costly. Natural gas is a bridge technology for the first years of climate policy because it has similar characteristics to coal but lower emissions. Nuclear power becomes competitive thanks to the constraints imposed by climate policy.[6] However, after 2070, renewables (photovoltaic and wind) become the major power generation technology. Investments over the years contribute to lower the cost of renewables thanks to learning. Learning spills over across world countries and contributes to a global penetration of renewables.

Backstop electricity represents a technology that is capable of supplying base-load power with zero emissions. Initially the cost of the backstop technology is very high. Investments in specific R&D reduce the cost of backstop electricity and make this technology competitive after 2050. Once backstop generation capacity is added to the power system learning contributes with R&D investments to reduce costs. Also in this case, learning spills over to all countries and the backstop technology is adopted globally.

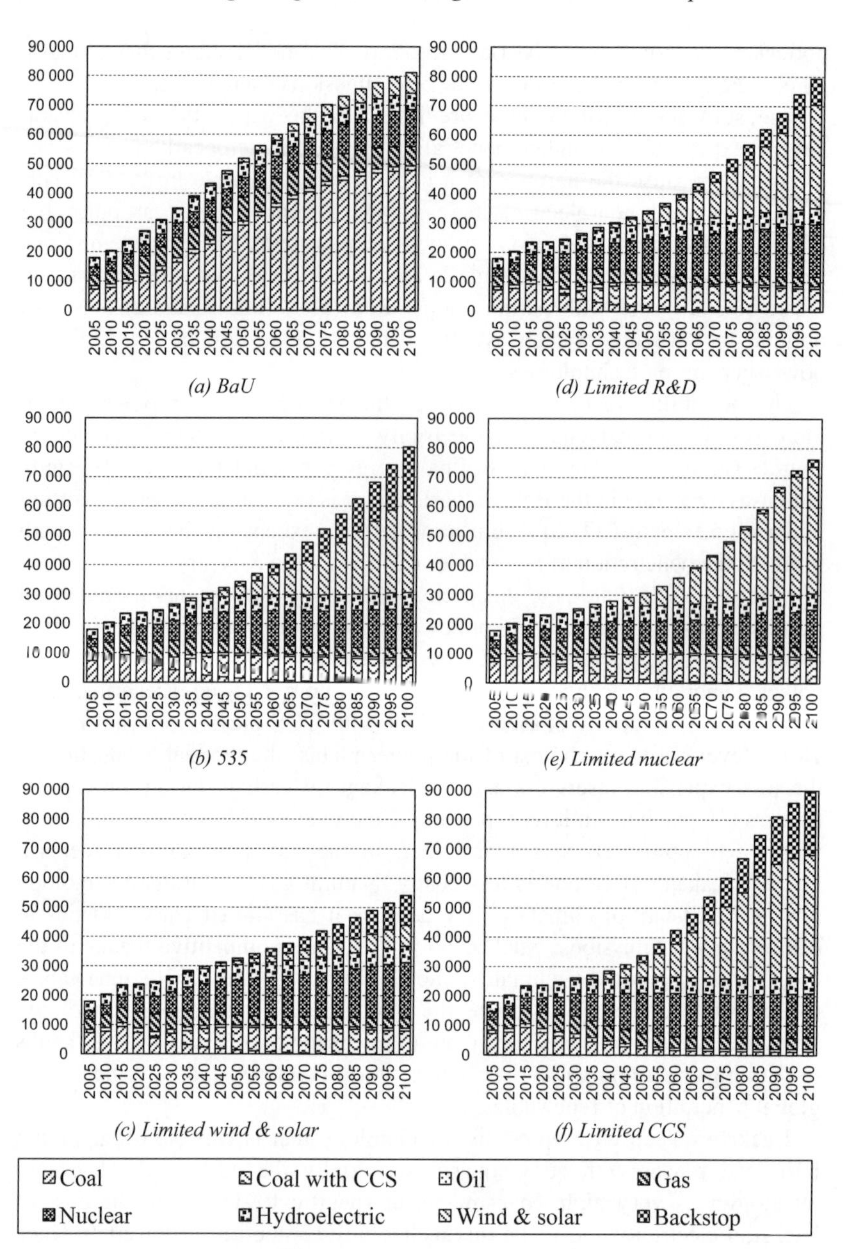

Figure 1.5 Electricity generation (TWH)

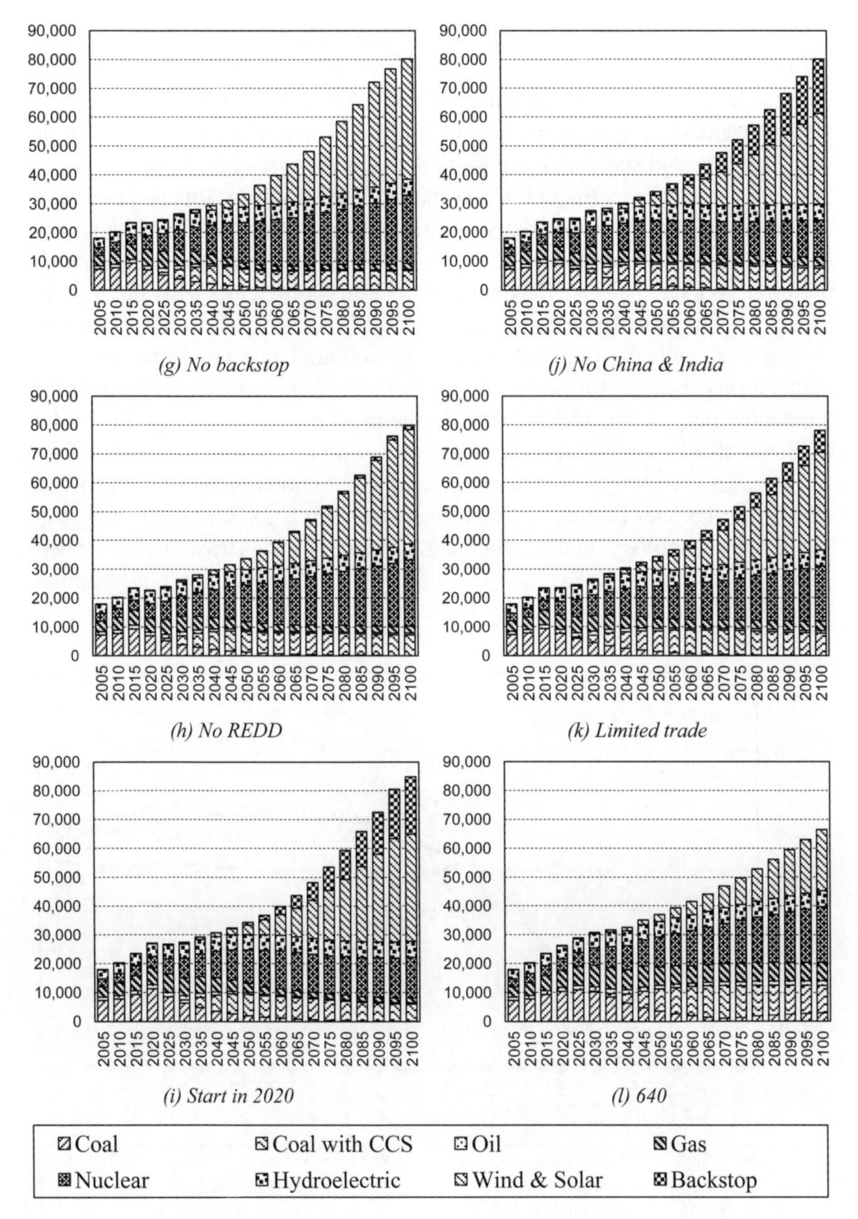

Figure 1.5 (Continued)

Figure 1.6 illustrates the role of CCS. The optimal amount of sequestered carbon is shown to be significant: about 6 $GtCO_2$/yr (about 1/4 of today's emissions) are stored underground in 2050. Over the entire century, about 400–460 $GtCO_2$ are injected into underground deposits, depending on the scenario. The 640 scenario has carbon deposits that are only about 10 percent larger than the more stringent 535 scenario. The reason for this is that a more stringent target calls for a relatively greater deployment of very low carbon technologies; renewable energies and nuclear power are thus progressively preferred to CCS, because they have lower emission factors. Advances in the capacity to capture CO_2 at the plant would increase CCS competitiveness; though this could be counterbalanced by potential leakage from reservoirs (simulations show that leakage rates of 0.5 percent per year would jeopardise the deployment of this technology). In the most recent version of the model, CCS can also be applied to natural gas power plants. Simulation results for a 550 ppm CO_2-eq target in 2100 show that it is optimal to first invest in coal with CCS and later (around 2050) also in gas with CCS. Gas with CCS is an important technology in the Middle-East and North Africa region, where about 30 percent of global investments in this technology are concentrated. Natural gas and coal compete for the scarce deposits for CO_2. Therefore gas

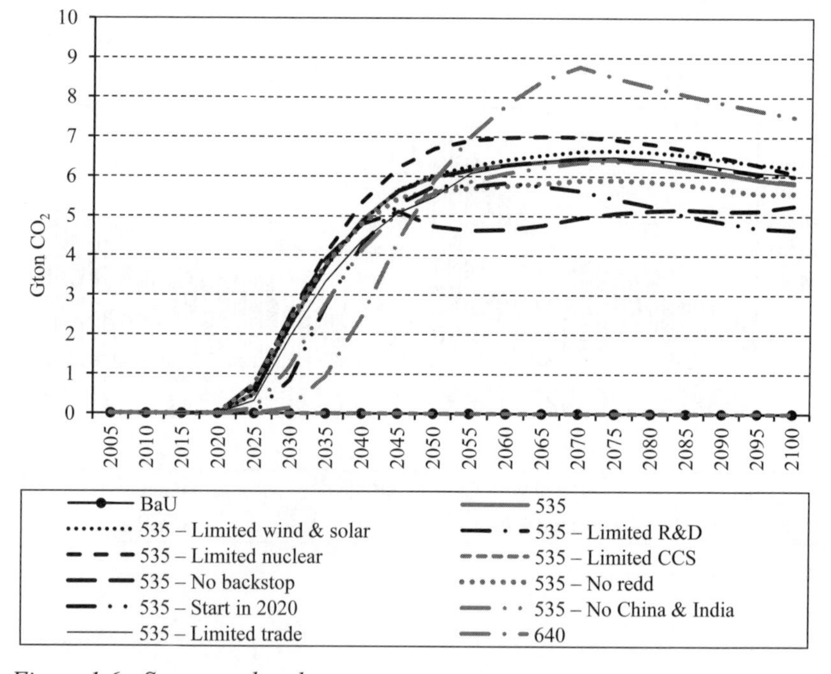

Figure 1.6 Sequestred carbon

with CCS crowds-out investments in coal with CCS. CCS is expected to play a major role in power plants fuelled by biomass. By storing underground emissions embodied in biomass it is possible to achieve net negative emissions. If adopted on a large scale, Biomass Energy with CCS (BECCS) has the potential to generate global net negative emissions, a pre-requisite to achieve the 2°C temperature target in 2100. Chapter 7 illustrates recent work with WITCH to examine the role of BECCS.

Figure 1.7 synthetically represents the total primary energy mix that results optimal under climate policy. Total primary energy includes energy used in the power sector and final energy uses. There are two important additions with respect to the electricity mix (Figure 1.5): final consumption of oil and gas and two carbon-free alternatives for oil in transport: biofuels and a generic backstop fuel. Final consumption of coal is included, but it is minimal and declining over time. All the scenarios show that it is optimal to tremendously reduce the consumption of oil in the transport sector. Natural gas and coal can still be used with CCS in a stabilisation scenario. Oil is instead used mostly in transport where it is not possible to capture emissions. The clear message from Figure 1.7 is that transport is a key sector for climate policy. Emission reductions in transport are, however, among the most expensive and will require tremendous technological and societal change. Traditional biofuels (biofuels in Figure 1.7) represent an alternative to oil but there are severe constraints to their expansion. Second-generation biofuels, which can be obtained using residues and widely available cellulosic feedstocks, may instead offer a viable alternative to oil. Also hydrogen or other fuels can potentially become viable alternatives for oil in transport. The generic carbon-free backstop-non electric is added in WITCH to describe a non-defined technology that has the potential to substitute oil in transport. Although much more expensive than oil at the beginning, the backstop fuel becomes competitive with investments in R&D. Also in this case, international knowledge spillovers distribute the cost reductions to all countries in the world and the backstop fuel becomes competitive at the global level after 2050. Recent work has added a detailed transport module to the WITCH model to study the optimal mix of fuel types and of car engines, including the hybrid and electric car. The transport module is described in Chapter 7.

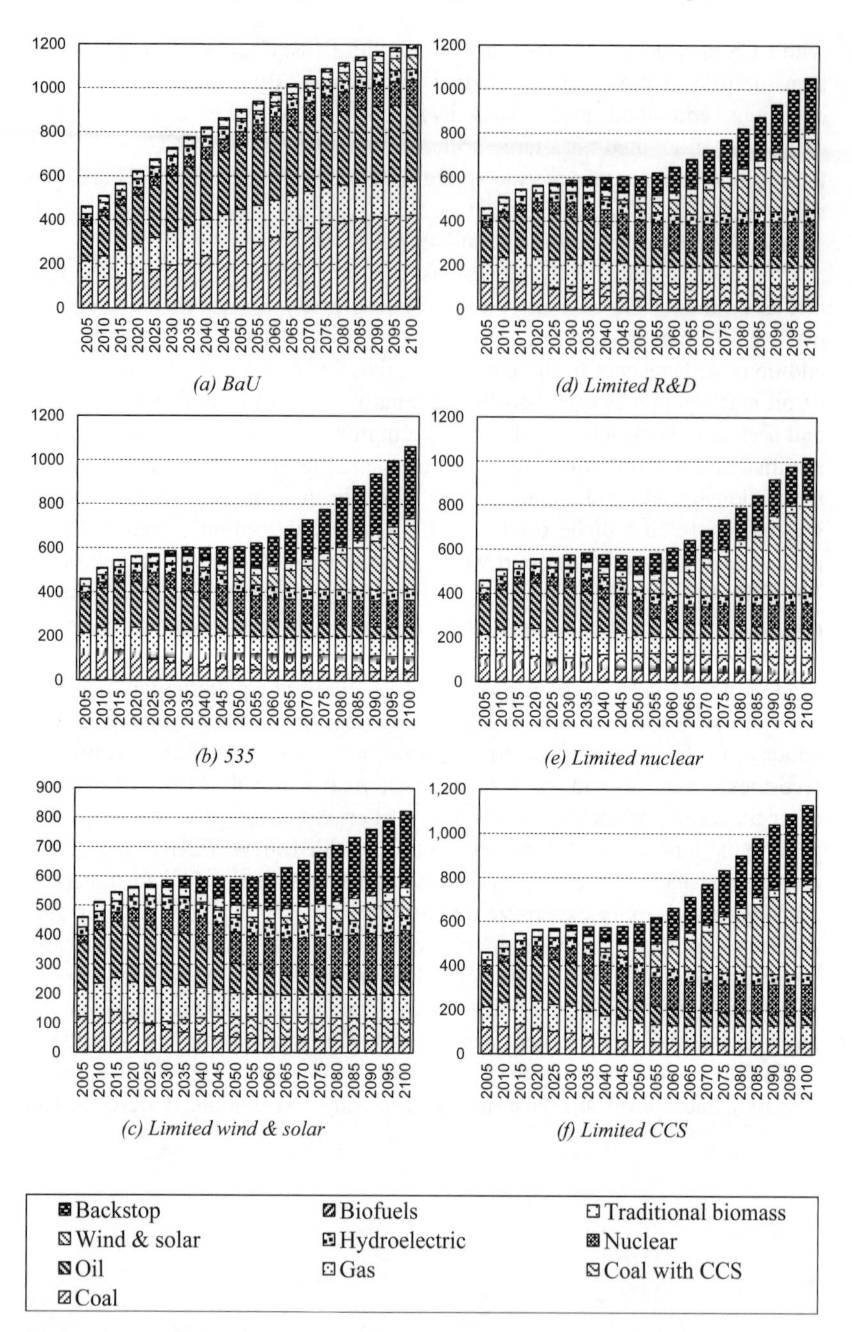

Figure 1.7 Total primary energy supply (EJ)

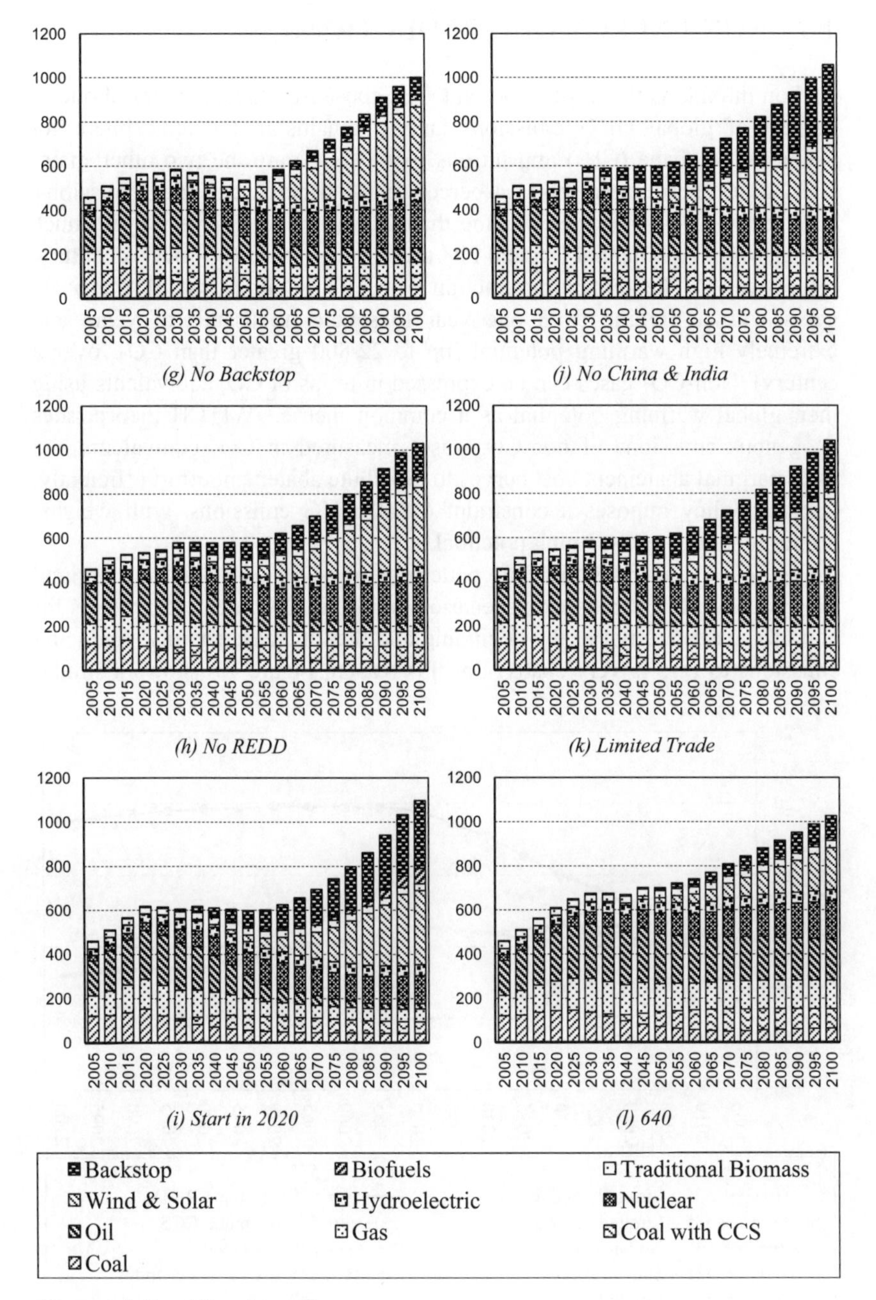

Figure 1.7 (Continued)

1.3 NON-ENERGY GHG REDUCTIONS

Carbon dioxide is the most important GHG because it accounts for about 77 percent of global GHG emissions and it remains in the atmosphere for centuries. Methane (CH_4) and nitrous oxide (N_2O) are the two other major GHG gases, accounting for 14.4 percent and 7.4 percent of total emissions. CH_4 and N_2O have shorter lifetime than CO_2 (12 and 114 years) but much higher warming potential (25 and 298 times greater than CO_2 over a century, respectively). The remaining fraction of GHG is composed of fluorinate gases with a lifetime from one year to several thousands of years and extremely high warming potential (up to 22 800 greater than CO_2, over a century). Non-CO_2 gases can be expressed in terms of CO_2 equivalents using their global warming potential as a common metric.[7] WITCH incorporates exogenous emissions of non-CO_2 gases measured in CO_2 equivalents and uses marginal abatement cost curves to distribute abatement effort efficiently. Climate policy imposes a constraint on all GHG emissions, with weights proportional to their warming potential.

Figure 1.8 illustrates the BaU pattern of non-CO_2 gases and the optimal trajectories in the stabilisation scenarios. Marginal abatement cost curves for non-CO_2 gases are flat at the beginning and then increase sharply. Reducing emissions to zero is very costly. For this reason, Figure 1.8 shows a similar

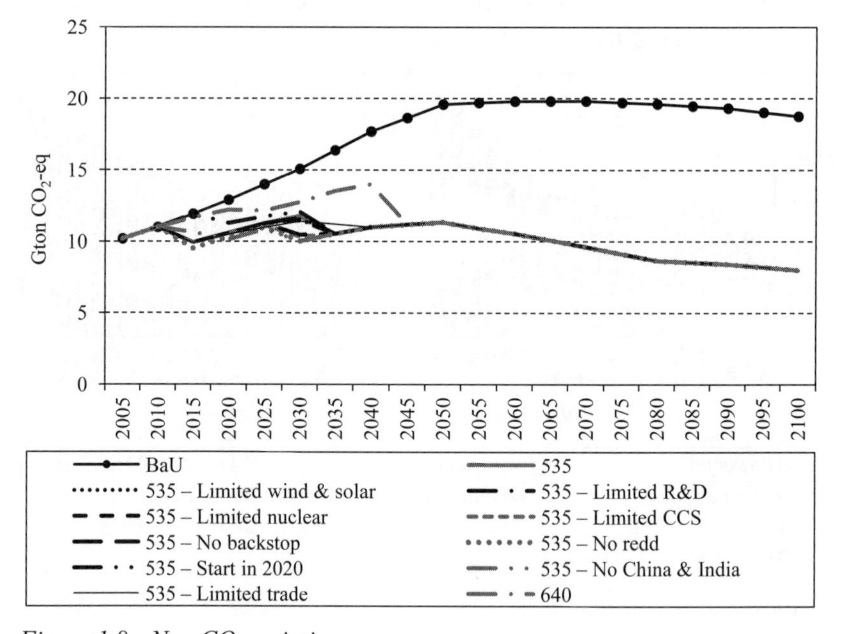

Figure 1.8 Non CO_2 emissions

pattern of emissions for all scenarios, including the 640 scenario: the model suggests that non-CO_2 mitigation actions should have the highest priority in all mitigation plans. Extending carbon pricing to non-CO_2 gases guarantees efficiency and should become a common attribute of climate policy.

Estimates of CO_2 emissions from Land Use and Land Use Change and Forestry (LULUCF) vary greatly in the literature. The IPCC AR4 estimates that CO_2 emissions from forestry were equal to 17.3 percent of total GHG emissions in 2004 (Rogner et al., 2007). In WITCH, the emissions from LULUCF are equal to 12.5 percent of the total emissions in 2005. A more sustainable management of forests and better soil management techniques can greatly contribute to the mitigation goal. WITCH includes an exogenous pattern of LULUCF emissions and marginal abatement cost curves. The model allocates resources to avoid emissions in the LULUCF sector so that marginal abatement costs are equal across all gases and all sectors. Emissions from LULUCF in the stabilisation scenarios are therefore endogenous and respond to the policy constraint as any other gas from any other sector. Since avoiding deforestation and promoting afforestation are relatively cheap mitigation measures (REDD), the model uses this option early, at full potential in all 535 ppm CO_2-eq scenarios (Figure 1.9). While REDD covers only 4 percent of cumulative abatement from 2010 to 2100 (Figure 1.10), it represents 15 percent of the abatement mix in 2025. The scenario with limited trade has the lowest amount of emission reductions from REDD because countries with large mitigation potential cannot sell emission credits and thus limit their abatement potential. In the 640 scenario, REDD becomes as attractive as in the 535 scenario after 2050. The model thus reveals that it would be optimal to direct resources soon to protect the world rainforests, to increase afforestation and to introduce new practices to manage carbon in soils (see also Carraro and Massetti, 2011a).

Figure 1.9 illustrates the optimal mix of mitigation actions divided in broad categories over the period 2010–2100. The optimal mix is stable even if the scenarios differ significantly. The 640 scenario shows a greater role for non-CO_2 gases, for the power sector and for REDD: emissions reductions in these sectors are less expensive than in final energy uses, notably in transport. Although important in the first years of climate policy, REDD gives only a marginal contribution in the long-run. Abatement of fossil fuels emissions is a necessary condition to achieve both mitigation targets.

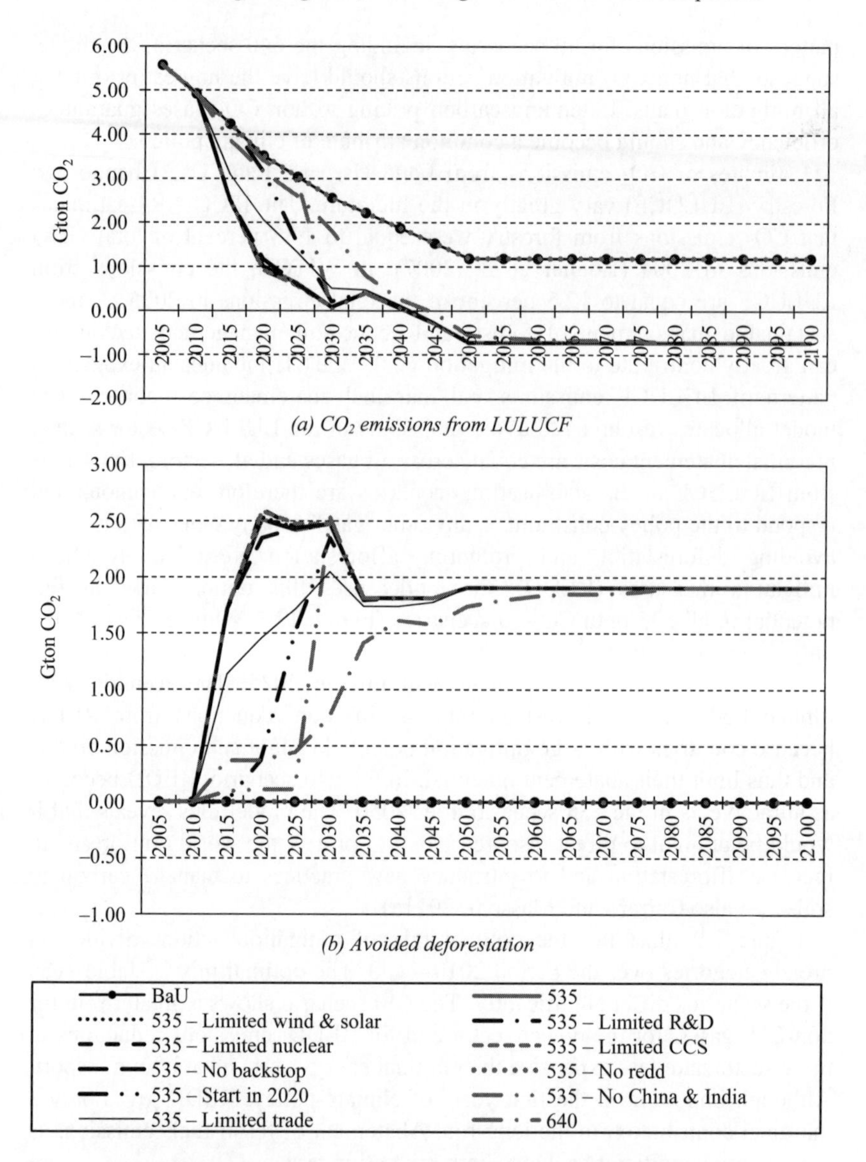

Figure 1.9 CO_2 emissions from LULUCF and avoided deforestation (REDD)

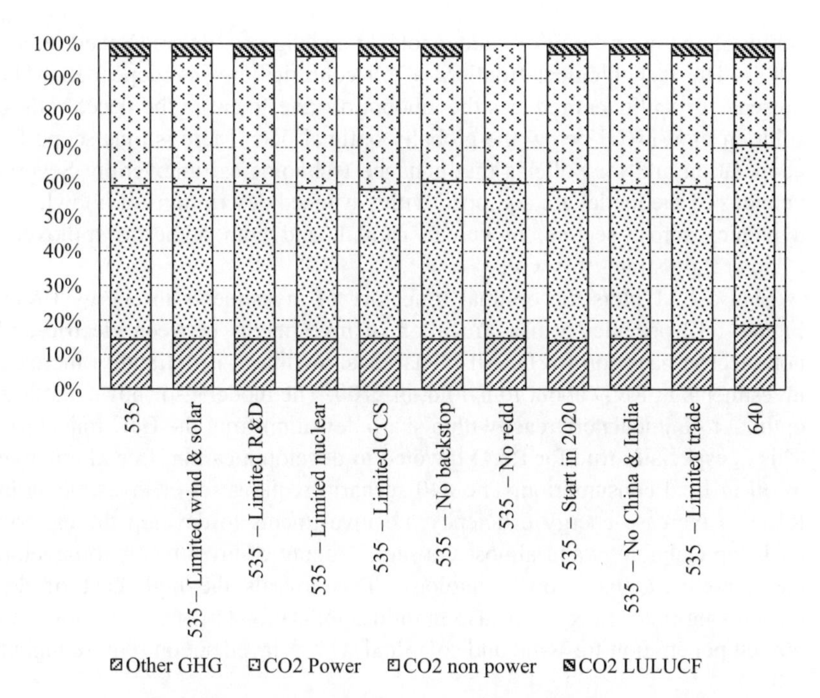

Figure 1.10 Distribution of cumulative abatement effort across broad sectors (2010–2100)

1.4 INNOVATION STRATEGIES FOR ENERGY EFFICIENCY AND TECHNOLOGY BREAKTHROUGH

WITCH has a rich description of endogenous knowledge dynamics. The model allows the study of the optimal combination of investments in currently available technologies and in R&D to generate technical progress both for energy efficiency and decarbonisation. WITCH features separate R&D investments for energy efficiency enhancements and for the development of breakthrough technologies in both the electric and non-electric sector. It is therefore possible to compute the equilibrium R&D investments that countries should implement to achieve the required improvements in energy efficiency and timely market penetration for new carbon-free energy technologies. These technologies are called backstops. They substitute nuclear power for power generation and oil in the non-electric sector. For a complete description, see the web site www.witchmodel.org.

Most importantly, knowledge spills over from one region to the other. Knowledge is a public good that can be appropriated (sometimes with some temporal lag) by those that have not contributed to create it. WITCH has a sophisticated description of endogenous knowledge spillovers (Bosetti et al., 2008). The technological frontiers of all countries are interdependent. The presence of spillovers reduces the amount of investment in the decentralised solution of WITCH below the globally optimal one. There is thus scope for cooperation and for policies that enhance technological innovation. Several versions of the model are equipped with advanced descriptions of knowledge dynamics across sectors, in human capital, and with domestic spillovers. Chapter 2 illustrates this work.

Figure 1.11 plots the optimal trajectory of investments in energy R&D. Figure 1.12 provides a disaggregation of investments between electric and non-electric backstop R&D. All scenarios show that it is optimal to increase investments in R&D about four-fold, in 2100. The model also shows that it is optimal to immediately react with a sharp deviation from the BaU trajectory. This is especially true for R&D devoted to develop a carbon-free alternative to oil in final consumption. The 640 scenario requires fewer investments in R&D to increase energy efficiency, no investments to develop the electric backstop technology and almost the same amount of investments to develop the non-electric backstop technology. This reveals the high cost of de-carbonising final energy uses also in mild stabilisation targets. Scenarios with limited penetration for wind and solar and with delayed action require higher

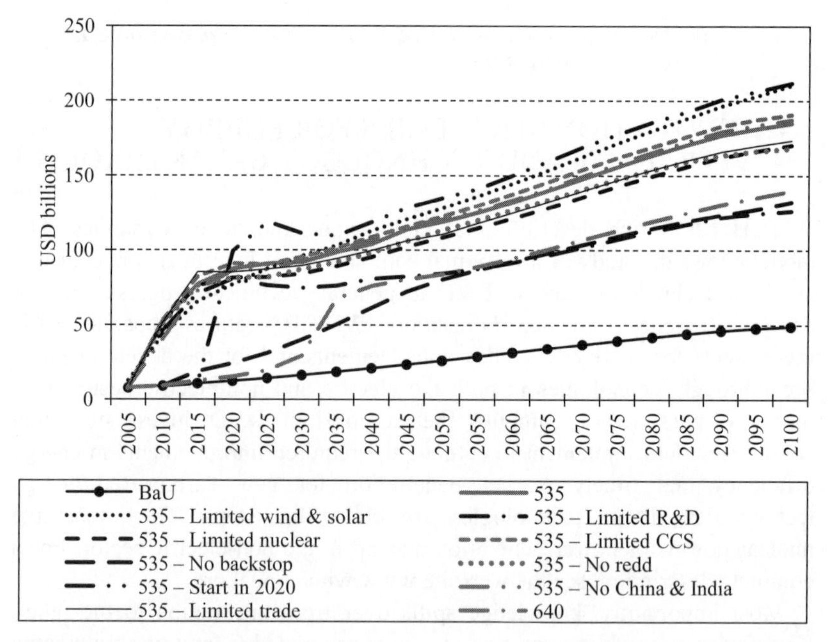

Figure 1.11 Energy R&D investments

investments in R&D. In particular, the start in 2020 scenario demands a dramatic increase of investments in backstop R&D in 2020. R&D to develop a backstop technology in the electric sector is high when there are limits to renewables. Investments in energy efficiency R&D follow a smoother path and are high when climate policy starts late and when there are limits to investments in the two backstop technologies. Figure 1.13 portrays the

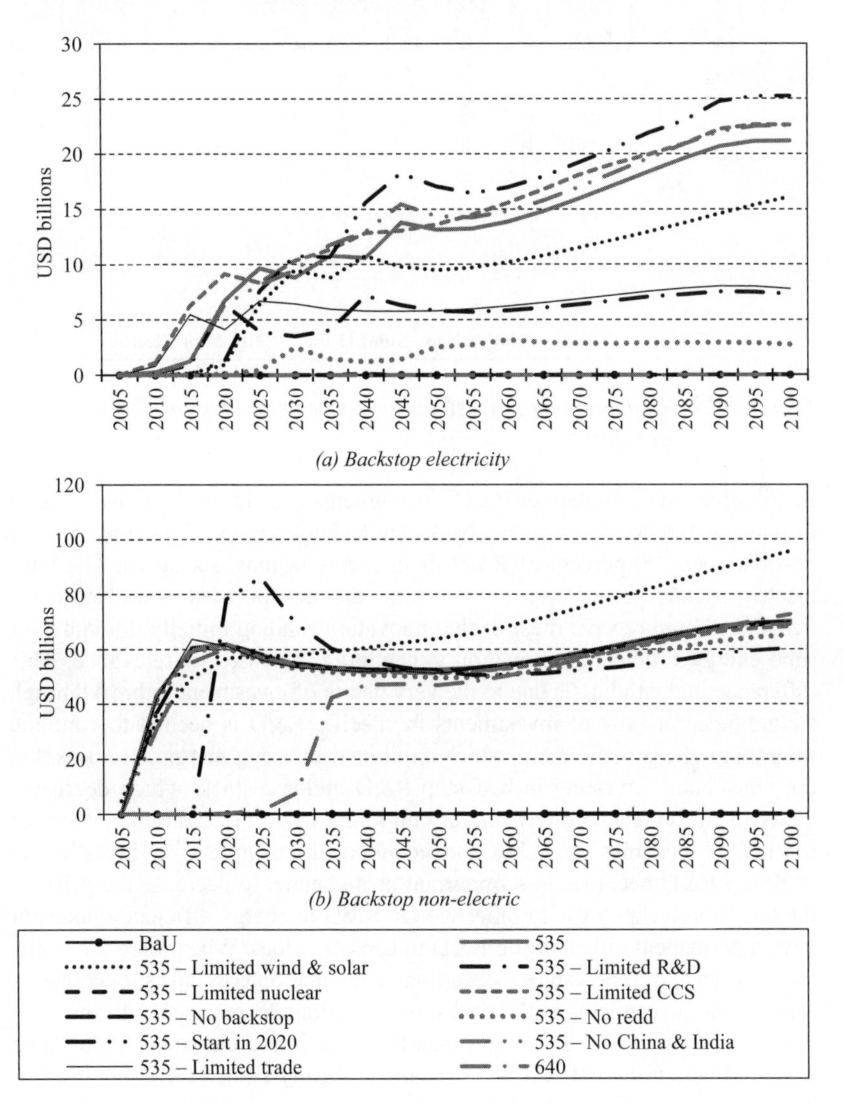

Figure 1.12 Investments in electric and non-electric backstop R&D

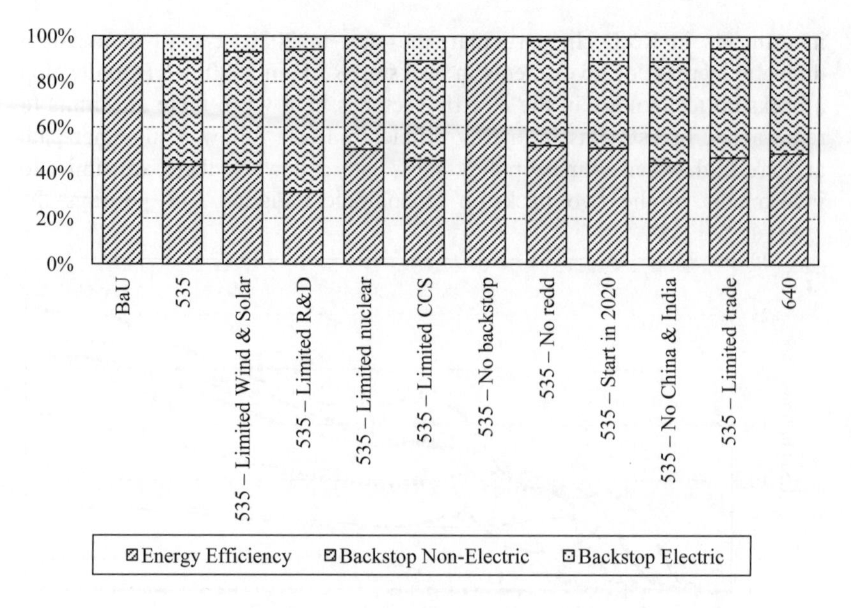

Figure 1.13 Distribution of investments in energy R&D across broad categories

distribution of cumulative R&D investments. R&D to increase energy efficiency and to develop the backstop technology for final energy uses absorbs about 90 percent of R&D investments in most scenarios. The non-electric sector, particularly to substitute the transport-led non-electric oil demand, should receive most of the innovation funding initially, though over time energy efficiency innovation expenditure increases its relevance. This difference in the timing is due to the very nature of investment in breakthrough technologies: a flow of investments in specific R&D is needed to continue improving energy efficiency, which exhibits decreasing marginal returns. On the other hand, investing in backstop R&D builds a stock, which decreases the costs of the technology with very high returns at the beginning. Once the technology becomes available and economically competitive, investing in backstop R&D becomes less important as a channel to decrease the price of the backstop technology. In other words, R&D in energy efficiency does not have a permanent effect, while R&D in backstop does. When there are limits to nuclear and renewables, decarbonising final energy uses becomes a priority and investments in the backstop technology increase even further.

The possibility of technology breakthroughs in the electricity sector also has an effect on the optimal investments in already known technologies. For example, the presence of backstop technologies greatly affects investments in renewables and in nuclear.

1.5 THE COST OF CLIMATE POLICY

The previous sections have stressed the importance of drastic changes in consumption and production of energy. They highlight the need to mobilise substantial resources towards carbon-free technologies and to change the equilibrium mix of energy, capital, and labour. The standard assumption in the class of optimal growth models – to which WITCH belongs – is that resources are efficiently allocated and a deviation from the equilibrium reduces the productivity of the economy and aggregate output. Of course, the cost of reducing emissions is compensated by reduced climate change impacts. Ideally, the stabilisation target would be fixed so that costs are equal to benefits. However uncertainties on the distribution and the magnitude of climate change impacts has pushed policy makers to follow a precautionary approach and to explore a set of 'safe' stabilisation targets. For this reason, this chapter does not assess the benefit of reducing emissions and does not factor it into the cost of climate policy. The purpose is rather to find the least-cost mix of investments that yield a desired stabilisation target. For each target and for each scenario it is possible to calculate the cost of achieving that target by comparing the discounted level of output in the BaU scenario and in the policy scenario.

WITCH's sophisticated setting includes several externalities at the global level (Chapter 2 illustrates a version of the model with also domestic externalities). First and foremost, WITCH includes pervasive knowledge externalities (both as learning-by-doing and learning-by-researching). Second, regions do not follow a globally optimal pattern of extraction of exhaustible resources because they do not internalise the effect of consumption on the global price of those resources. Therefore exhaustible resources are harvested at a rate higher than the socially optimal one. Since climate policy fosters innovation and reduces the use of exhaustible resources it also increases global welfare. However, the gains are minimal and climate policy is costly (without including the environmental benefit).

Figure 1.13 shows net present value losses of Gross World Product (GWP) for both climate policy targets and different technology settings. Costs are highly non-linear in the abatement effort: from 0.9 percent of GWP in the 640 ppm CO_2-eq scenario, they increase to 2.8 percent of GWP in the 535 ppm CO_2-eq scenario. The cost difference between the two mitigation policies is a direct consequence of the different scale of the required energy sector modifications. It also stems from the non-linearity of endogenous marginal abatement curves in the model. The 535 ppm CO_2-eq policy requires drastic cuts in emissions, especially in the second half of the century (see Figure 1.2) and this increases costs considerably. Results with the latest version of WITCH show that BECCS allow to reduce the cost of achieving

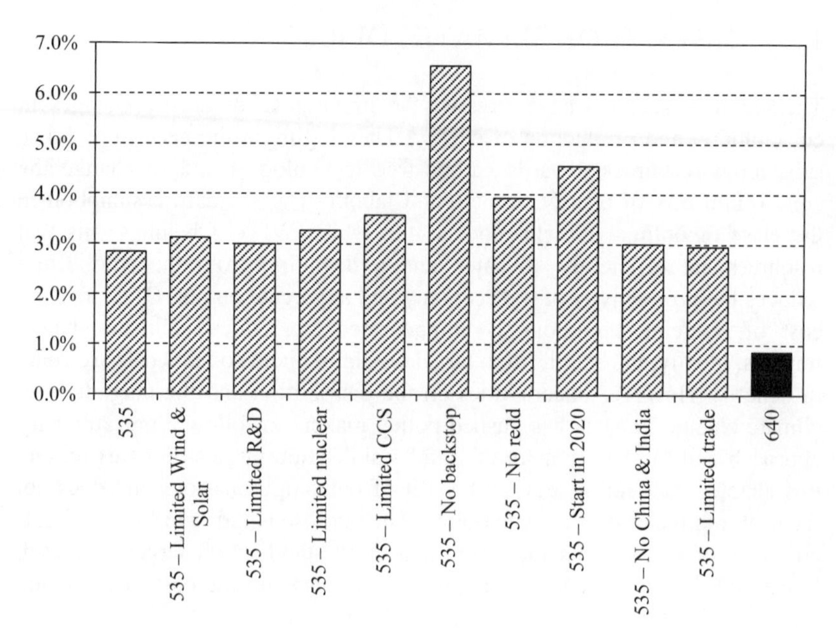

Figure 1.14 The cost of climate mitigation policy

the 550 ppm CO_2-eq target. However, achieving the 450 ppm CO_2-eq target costs about 4 percent of GWP, thus confirming the very steep marginal abatement cost curves of previous studies (Massetti, 2011; Carraro and Massetti 2012; Massetti and Tavoni, 2012; Carraro et al., 2012; Bosetti et. al., 2012).

Figure 1.14 also shows the economic effect of limiting technological and policy options. Limits to renewables, R&D for energy efficiency, and nuclear do not raise costs considerably. When only one option is not available the model has enough flexibility to find substitutes in the same cost category. Applying more than one limit would increase costs considerably (see Luderer et al., 2012). Limits to CCS increase the cost of achieving the 535 ppm CO_2-eq target by 25 percent. Without backstop technologies costs increase by 130 percent, up to 6.6 percent of GWP, an astonishing figure, which reveals the importance of expanding the current set of tools to fix the global warming problem. Also limiting inexpensive abatement options in developing countries using REDD has remarkable high costs (3.9 percent of GWP). Delaying global action by only a decade requires much deeper emissions reductions in the future. Since marginal abatement costs are highly non-linear, global costs increase by 60 percent.

In all the scenarios under examination, the policy tool is a global cap-and-trade system. Emission rights are allocated to world regions and can be traded internationally. Banking and borrowing are not allowed. Figure 1.15 reports the endogenous price of emission permits. The 535 ppm CO_2-eq stabilisation target requires a price on GHG emissions much higher than the 640 ppm CO_2-eq policy scenario. The price spikes when REDD is not available and when there are limits to backstop technologies. Slow action to achieve a global agreement on climate change produces a very high price at the end of the century, when very aggressive action to reduce emissions is needed to compensate for the initial delay. Interestingly, if China and India are not included in a global deal on climate change until 2030, the price of GHG emissions will be very high in the first period, then it will decline when developed economies have access to carbon credits from China and India, and then it will start to grow again.

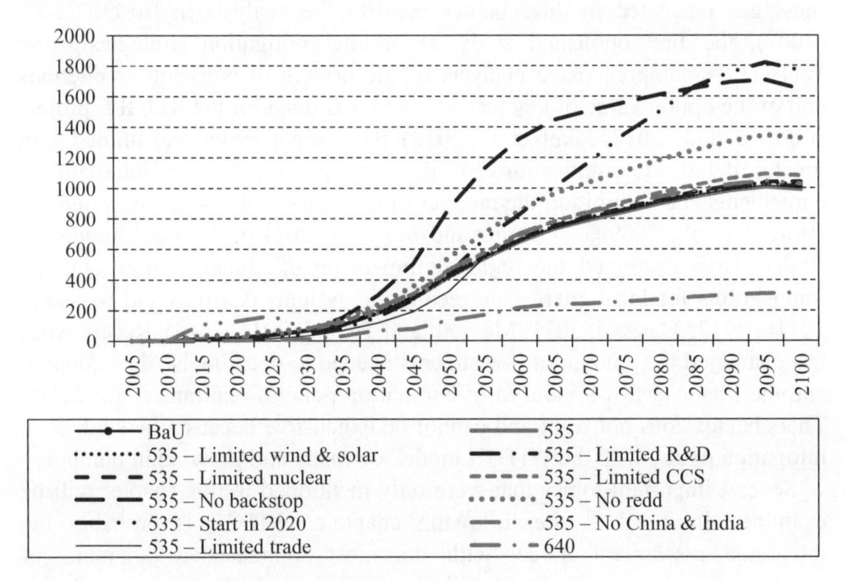

Figure 1.15 The carbon price (USD/tCO_2-eq)

1.6 CONCLUSIONS

This chapter examined optimal investment strategies in the energy sector and optimal mitigation strategies to reduce non-CO_2 gases and emissions from LULUCF for two climate policy targets. Results show that stabilising GHG concentrations at 640 ppm CO_2-eq in 2100 is feasible at reasonable economic costs. Reducing the level of concentrations by an additional 105 ppm CO_2-eq, costs at least 2.8 percent of GWP. If some carbon-free or low-emissions power technologies have penetration limits, if a backstop carbon-free technology for final energy uses is not available, if there are delays or incomplete international participation in the mitigation effort, costs will increase, also substantially.

Both energy efficiency and the decarbonisation of energy should be pursued. Currently known technologies in the power sector such as nuclear, renewables, and CCS will play a major role. At the same time, R&D investments for the development of new technologies, especially in the transport sector, will be required. Public R&D expenditures should increase considerably, over the peak levels of the 1980s for at least three decades. Given the long time lags inherent to the innovation process, such investments should be made soon.

Although the model has evolved considerably over the years, the key messages presented in this chapter confirm the analysis in Bosetti et al. (2007), the first published study of optimal mitigation strategies using WITCH. A comprehensive analysis of the drivers of emissions reductions and of the option value of key technologies was done for the RECIPE project (Luderer et al., 2012; Jakob et al., 2012). Several papers have examined with greater detail the implications of delaying participation in international agreements and/or limiting the number of countries that engage in mitigation (Bosetti et al., 2009b,c; Carraro and Massetti, 2011a). A large number of studies have examined the regional dimension of climate policy and the implications for local energy and economic systems (Carraro and Massetti, 2011b, 2012; Massetti, 2011; Massetti and Tavoni, 2011, 2012). Recent work has examined the amount of investments needed to decarbonise the economy and the financial flows induced by mitigation policies (Carraro et al., 2012). This chapter does not want and cannot be exhaustive because the studies on mitigation policy with the WITCH model are many and growing in number.

Several important topics that were only mentioned in this chapter will be examined in detail in the following chapters. WITCH is probably the integrated assessment model with the most sophisticated dynamics of technological change. Chapter 2 will illustrate a large body of research in this area.

The analysis in this chapter has generally assumed that all countries agree to start reducing emissions immediately. Is this a realistic assumption? Is the large global coalition stable? Will all the partners have the incentive to form it and not to deviate from the declared objective? Several papers have exploited WITCH's game-theoretic set-up to explore the main drivers that lead countries to cooperate or not cooperate to control global warming. Chapter 3 collects the most important findings in this research area.

This chapter has highlighted the importance of investing in R&D, especially to develop breakthrough technologies assuming that the return to R&D investments is deterministic. However, there is nothing more uncertain than investments in research at the frontier. Also, the analysis until now, has assumed that policy makers are able to convey a convincing signal on when and by what amount emissions will start to drop. Unfortunately, this is definitely not true. Chapter 4 examines how uncertainty affects the optimal pattern of investments, also using a specially-built stochastic version of WITCH.

The role of REDD in contributing to the overall mitigation goal is examined in Chapter 5. The analysis shows how sensitive the scenarios are to different estimates of present and future emissions from LULUCF.

While this chapter has examined the least-cost options to achieve a given stabilisation target, Chapter 6 derives the optimal level of concentrations in 2100 by comparing the cost of mitigation, the benefit from reducing the impacts of climate change, and the cost of adapting to climate change. While mitigation reduces the overall threat of climate change by containing the increase of the global average temperature, adaptation minimises (maximises) negative (positive) impacts at the local level. The exact mix of mitigation and adaptation is a new challenging research area.

Finally, Chapter 7 illustrates new developments of the WITCH model, especially the possibility to use biomass with CCS to obtain global net negative emissions, a pre-requisite to achieve very low stabilisation targets.

NOTES

1. This chapter is based on a set of scenarios developed using the same version of the WITCH model. The scenarios are available at the website www.witchmodel.org. Short sections of this chapter are abstracts from the paper Bosetti et al. (2009c). The model is described in Bosetti et al. 2006, 2007, 2009d and in the website www.witchmodel.org/.
2. Major Economies Forum on Energy and Climate: http://www.majoreconomiesforum.org/.
3. The so called '20–20–20' directive also includes a mandatory target for expanding the share of renewable energy to a minimum of 20 percent of total primary energy supply and to increase energy efficiency by 20 percent.
4. For an analysis of a market with banking see Bosetti et al. (2009a).
5. In this book a backstop technology is defined as a zero-emission large-scale energy technology. Two backstop technologies are modeled in WITCH: one in the power sector and the other for final energy use.
6. Recent developments in the extraction of shale gas have radically changed natural gas markets and suggest that natural gas might have a bigger role to play in short- and mid-term mitigation policy.
7. All data on GHG is from Forster et al. (2007).

REFERENCES

Bosetti, V., C. Carraro and E. Massetti (2009a), 'Banking permits: Economic efficiency and distributional effects', *Journal of Policy Modeling*, **31**(3), 382–403.

Bosetti, V., C. Carraro, E. Massetti, A. Sgobbi and M. Tavoni (2009c), 'Optimal energy investment and R&D strategies to stabilise greenhouse gas atmospheric concentrations', *Resource and Energy Economics*, **31**(2), 123–137.

Bosetti, V., C. Carraro, E. Massetti and M. Tavoni (2008), 'International energy R&D spillovers and the economics of greenhouse gas atmospheric stabilization', *Energy Economics*, **30**(6), 2912–2929.

Bosetti, V., C. Carraro, A. Sgobbi and M. Tavoni (2009c), 'Delayed action and uncertain stabilisation targets. How much will the delay cost?', *Climatic Change*, **96**(3), special issue on 'The economics of climate change: Targets and technologies', 299–312.

Bosetti, V., C. Carraro, M. Galeotti, E. Massetti and M. Tavoni (2006), 'WITCH: A world induced technical change hybrid model', *The Energy Journal*, special issue on 'Hybrid modeling of energy-environment policies: Reconciling bottom-up and top-down', 13–38.

Bosetti, V., C. Carraro and M. Tavoni (2009b), 'Climate change mitigation strategies in fast-growing countries: The benefits of early action', *Energy Economics*, **31**(S2), S144–S151.

Bosetti, V., C. Carraro and M. Tavoni (2012), 'Timing of mitigation and technology availability in achieving a low-carbon world', *Environmental and Resource Economics*, **51**(3), 353–369.

Bosetti, V., E. De Cian, A. Sgobbi and M. Tavoni (2009d), 'The 2008 WITCH model: New model features and baseline', Nota di Lavoro 85.2009, Milan, Italy: Fondazione Eni Enrico Mattei.

Bosetti, V., E. Massetti and M. Tavoni (2007), 'The WITCH model: Structure, baseline and solutions', Nota di Lavoro 10.2007, Milan: Fondazione Eni Enrico Mattei.

Carraro, C., A. Favero and E. Massetti (2012), 'Investments and public finance in a green, low carbon economy', *Energy Economics*, **34**(S1), S15–S28.

Carraro, C. and E. Massetti (2011a), 'Beyond Copenhagen: A realistic climate policy in a fragmented world', *Climatic Change*, **110**(3), 523–542.

Carraro, C. and E. Massetti (2011b), 'Editorial', *International Environmental Agreements, Law, Economics and Politics*, special issue on 'Reconciling domestic energy needs and global climate policy: Challenges and opportunities for China and India', **11**(3), 205–208.

Carraro, C. and E. Massetti (2012), 'Energy and climate change in China', *Environment and Development Economics*, **17**(6), 689–713.

Clarke, L., J. Edmonds, V. Krey, R. Richels, S. Rose and M. Tavoni (2009), 'International climate policy architectures: Overview of the EMF 22 international scenarios', *Energy Economics*, **31**(2), 64–81.

Forster, P., V. Ramaswamy, P. Artaxo, T. Berntsen, R. Betts, D.W. Fahey, J. Haywood, J. Lean, D.C. Lowe, G. Myhre, J. Nganga, R. Prinn, G. Raga, M. Schulz and R. Van Dorland (2007), 'Changes in atmospheric constituents and in radiative forcing', in IPCC (Intergovernmental Panel on Climate Change), *Climate Change 2007: The Physical Science Basis. Contribution of Working Group I to the Fourth Assessment Report of the Intergovernmental Panel on Climate Change* [S. Solomon, D. Qin, M. Manning, Z. Chen, M. Marquis, K.B. Averyt, M.Tignor and H.L. Miller (eds)], Cambridge, UK and New York, NY, USA: Cambridge University Press, pp. 131–234.

Höhne, N., C. Taylor, R. Elias, M. den Elzen, K. Riahi, C. Chen, J. Rogelj, G. Grassi, F. Wagner, K. Levin and E. Massetti (2012), 'National greenhouse gas emissions reduction pledges and 2°C – Comparison of studies', *Climate Policy*, **12**(3), 356–377.

IPCC (Intergovernmental Panel on Climate Change) (2007), *Climate Change 2007: Synthesis Report*, Geneva: IPCC.

Jakob, M., G. Luderer, J. Steckel, M. Tavoni and S. Monjon (2012), 'Time to act now? Assessing the costs of delaying climate measures and benefits of early action', *Climatic Change*, special issue on 'The economics of decarbonization in an imperfect world', **114**(1), 79–99.

Luderer, G., V. Bosetti, M. Jakob, M. Leimbach, J.C. Steckel, H. Waisman and O. Edenhofer (2012), 'The economics of decarbonizing the energy system – Results and insights from the RECIPE model intercomparison', *Climatic Change*, special issue on 'The economics of decarbonization in an imperfect world', **114**(1), 9–37.

Massetti, E. (2011), 'Carbon tax scenarios for China and India: Exploring politically feasible mitigation goals', *International Environmental Agreements, Law, Economics and Politics*, special issue on 'Reconciling domestic energy needs and global climate policy: challenges and opportunities for China and India', **11**(3), 209–227.

Massetti, E. and M. Tavoni (2011), 'The cost of climate change mitigation policy in Eastern Europe, Caucasus and Central Asia', *Climate Change Economics*, **2**(4), 341–370.

Massetti, E. and M. Tavoni (2012), 'A developing Asia emission trading scheme (Asia ETS)', *Energy Economics*, **34**(S3), S436–S443.

Rogner, H.-H., D. Zhou, R. Bradley, P. Crabbé, O. Edenhofer, B. Hare (Australia), L. Kuijpers and M. Yamaguchi (2007), 'Introduction', in IPCC (Intergovernmental Panel on Climate Change), *Climate Change 2007: Mitigation. Contribution of Working Group III to the Fourth Assessment Report of the Intergovernmental Panel on Climate Change* [B. Metz, O.R. Davidson, P.R. Bosch, R. Dave, L.A. Meyer (eds)], Cambridge University Press, Cambridge, UK and New York, NY, USA, pp. 97–116.

UNEP (United Nations Environment Programme) (2010), 'The emissions gap report – Are the Copenhagen accord pledges sufficient to limit global warming to 2°C or 1.5°C? A preliminary assessment', United Nations Environment Programme, November 2010.

2. Shifting the Boundary: The Role of Innovation

Lea Nicita

It's widely recognised that tackling climate change at bearable social costs will involve far-reaching technological changes, especially in the energy sector. Chapter 1 has shown that even moderate climate mitigation targets ask for a dramatic shift in the ways that energy is supplied and used.[1] Greater energy efficiency and widespread deployment of low-carbon technologies are both needed. Chapter 1 has also shown that it is optimal to massively increase investments in innovation in order to boost energy efficiency and to decarbonise the energy sector. For this reason technical change has become one of the key dimensions of the economics of climate change and policy modelling.

During the last decade the description of technical change in integrated models for climate policy has greatly improved (Gillingham et al., 2008; Carraro et al., 2010). Understanding and accurately characterising the process of technical change is a challenging task. Technical change responds to changes in economic incentives, or in policy regimes and regulations. Whereas the former process is referred to as endogenous technical change, the latter phenomenon is known as induced technical change (ITC). The majority of climate economy models show that induced technical change substantially changes the long-run costs of climate policy and the optimal timing of mitigation.

Several integrated assessment models describe endogenous technical change in clean technologies, but most neglect innovation in other sectors. The effect of mitigation policies on the direction and the pace of technical change is therefore still a matter of debate. Some climate-economy models generate scenarios of research and development (R&D) spending or learning, but most cannot say if climate policy will increase of decrease economy-wide R&D investments. As a consequence, climate-economy models rarely represent knowledge externalities and examine the interplay between R&D and climate mitigation policies.

The theoretical argument which postulates the presence of market failures in the R&D sector is supported by the empirical evidence. Many studies have shown that the social rate of return on R&D expenditure is higher than the private one (Griliches, 1957, 1992; Mansfield, 1977, 1996; Jaffe, 1986; Hall, 1996; Jones and Williams, 1998). Those studies show that the marginal social rate of return to R&D investments ranges between 30 and 50 percent while the private return to R&D investments ranges between 7 and 15 percent. For this reason, the new growth theory has emphasised the importance of international R&D knowledge spillovers (Grossman and Helpman, 1991, Chapters 11 and 12), and of both intrasectoral and intersectoral R&D knowledge spillovers in explaining observed differences in countries' productivity (Jones, 1999; Li, 2000).

Externalities plague all R&D activities, but when it comes to technologies for carbon emissions reduction, the difference between the private and the social rate of return to R&D investment is higher because of two market failures (Nordhaus, 1991). First, without a price on carbon that equates the global and the private cost of emitting greenhouse gases (GHG), all low emissions technologies are relatively disadvantaged and the level of investment is therefore sub-optimal. Second, the private return to investment in R&D is lower than the social return of investment due to the incomplete appropriability of knowledge creation, thus pushing investments further away from the socially optimal level.

This chapter builds on Chapter 1 and examines two sets of major policy questions regarding climate policy and innovation using advanced versions of WITCH that model the direction of technical change and knowledge externalities.

The first set of questions concerns the direction of technical change between the energy sector and non-energy sectors. The main research questions are the following: will a low-carbon economy be more likely to have a higher or a lower rate of technological innovation and human capital formation than an economy based on fossil fuels? And, would estimates of climate policy costs change in the presence of a detailed and articulated specification of endogenous technical change?

The second set of questions concerns knowledge externalities. The main research questions are the following: Can we expect that the stabilisation policy will drive the economies closer to or farther from the socially optimal level of innovation? What is the optimal mix of policies to support the optimal level of innovation, both in 'green' and 'dirty' technologies?

Various types of public policies may influence the rate and direction of technological change, including carbon pricing, R&D policies and subsidies to the deployment of existing technologies. These are explored here, with the purpose of assessing the effects of alternative policy mixes on endogenous technical change in the energy sector and the entire economy, and on the costs of achieving given GHG concentration stabilisation targets.

The chapter is organised as follows. Section 2.1 illustrates how technical change is modelled in the standard version of WITCH, used in Chapter 1 and throughout this book. The standard version of the model includes endogenous R&D investments in energy efficiency and in two energy backstop technologies. International R&D spillovers in the energy sector are endogenous in the standard version of the model. Section 2.2 explores issues related to some international dimensions of technical change, such as knowledge spillovers, global R&D policies and cooperation for technology deployment. Sections 2.3 and 2.4 illustrate several versions of the model that incorporate a more advanced treatment of technical change and address the two broad research questions introduced above. Section 2.3 studies the implications for climate policy of the presence of knowledge externalities and the different effects of climate policy on economy-wide innovation versus energy and carbon-saving innovation. Section 2.4 introduces the enhanced version of the WITCH model embodying R&D accumulation and human capital formation. The two models are used to illustrate the optimal mix of climate, knowledge and education policies. Concluding remarks follow.

2.1 TECHNICAL CHANGE IN THE ENERGY SECTOR

As shown in Chapter 1 reducing CO_2 emissions eventually depends on the uptake of energy-efficient and low-carbon technologies. Reaching ambitious concentration targets is challenging and will require significant investment and a reallocation of financial resources towards energy related innovation. A portfolio of low-carbon technologies as well as increasing energy efficiency will be needed to sustain emissions reductions.

Most emerging low-carbon technologies are not yet available or have higher costs than fossil-fuels, The empirical evidence shows that these technologies are subject to substantial 'learning effects', for instance, their cost decreases as experience accumulates through cumulative production. Reducing their costs and speeding up their diffusion will require increased expenditure at all stages of the technology development process to make these technologies competitive at market prices.

A distinguishing feature of WITCH is that it embodies different stages of technological change (innovation, deployment and diffusion) and both incremental and radical technology advances (see www.witchmodel.org for a detailed description).

Technical change in WITCH is both autonomous and endogenous. The autonomous component consists of a neutral technical change increasing the total factor productivity of each region, at declining rate.

In the basic set up of the model, endogenous technical change occurs in the energy sector and is driven by learning by researching and learning by doing processes. For details see Bosetti et al. (2006, 2007, 2009a).

R&D can be directed towards building three different stocks of knowledge creation. One stock increases the energy efficiency of the economy, another stock reduces the cost of a backstop technology for the power sector and a third stock reduces the cost of a backstop technology for the non-energy sector. Backstop technologies provide energy without GHG emissions, in virtually infinite amount. They are used to provide a compact representation of an unknown portfolio of advanced technologies that will become commercially viable if there are sufficient investments in research (e.g. fuel cells, advanced biofuels, advanced nuclear technologies etc.).

All knowledge stocks follow the same dynamics. At each point in time, new knowledge arises from domestic R&D investments through an innovation possibility frontier characterised by intrasectoral and international knowledge spillovers. In each region new ideas stem from the domestic stock of energy knowledge, accumulated over time in the sector, and from knowledge externalities from the international pool of sector-specific knowledge. For example, investments in non-electric backstop R&D in Europe spill over to other regions by increasing the productivity of local non-electric backstop R&D investments.

Knowledge spillovers are not totally free nor are they the same for all regions of the world (for more detail see Bosetti et al., 2008). The contribution of foreign spillovers to domestic knowledge accumulation depends on each region's distance from the technology frontier, measured as the difference between each region's own knowledge capital stock and the knowledge stock built in high-income regions, and on its absorptive capacity, measured as the ratio of a region's knowledge capital to the technological frontier.

High income countries, even if close to technology frontier, take advantage of knowledge spillovers because knowledge is assumed to be heterogeneous across countries. High-income countries have a high absorptive capacity and get the most from international knowledge spillovers. Knowledge externalities are potentially very large for low-income countries because they are very far from the technology frontier. However the effective amount of knowledge that can be used to increase domestic productivity of research is limited by a low absorption capacity. Countries that are further away from the technological frontier are less capable to absorb knowledge from the international knowledge pool because they lack laboratories, scientific bodies and researchers that make it possible to exploit foreign knowledge.

International learning spillovers reduce the cost of renewable power generation technologies. The cost of these technologies follows a one-factor learning curve and declines with the world installed capacity.

In line with the most recent literature, the emergence of backstop technologies is modelled through so-called 'two-factor learning curves', in which the cost of a given backstop technology declines both with investment in dedicated R&D and with technology diffusion because of learning-by-doing effects (see among others Kouvaritakis et al. 2000). This formulation is meant to overcome the limitations of single factor experience curves, in which the cost of a technology declines only through 'pure' learning-by-doing, without the need for R&D investment (Nemet, 2006). Nonetheless, modelling long-term and uncertain phenomena such as technological evolution is inherently difficult and calls for caution in interpreting the exact quantitative results and for sensitivity analysis, as shown in Bosetti et al. (2009). In WITCH, backstop technologies substitute linearly for nuclear power in the electric sector, and for oil in the non-electric sector. Once technologies become competitive their penetration is assumed to be gradual rather than immediate and complete. The upper limit on penetration is set at 5 percent of the total consumption in the previous period of technologies other than the backstop.

Investments in innovation play an important role in all climate policy scenarios studied with WITCH. While investments in energy efficiency R&D and the diffusion of renewable technologies are included in all mitigation portfolios generated by WITCH, many studies show that it is optimal to invest in backstop technologies only when concentrations of GHG in the atmosphere are forced to remain below 650 ppm CO_2-eq at the end of the century.

In a concentration target equal to 550 ppm CO_2-eq at the end of the century, R&D expenditures are found to quadruple with respect to the BaU and, as displayed in Figure 2.1, to rise up to 0.08 percent of global GDP. The time pattern of R&D spending reflects the nature of investment in breakthrough technologies. These are characterised by very high marginal returns at the beginning, which then decline gradually as the R&D stock is built up and the potential for further cost reductions, through additional R&D investments, fades out. The cost of the backstop technologies follows an inverted S-shaped path. R&D activities bring costs down rapidly in the early phases, when backstops is very expensive. After 2030-2040, further cost declines occur mainly through learning by doing, as the technologies are deployed.

These results rely on the assumption that mitigation policy is efficient worldwide and all mitigation technologies are available. Otherwise even a relatively modest mitigation target would generate an explicit or implicit high price of carbon in some countries and thus a strong incentive to provide backstop technologies to de-carbonise the energy sector.

De Cian et al. (2011) study the case in which global climate policy imposes a higher price on emissions in developed countries than in developing countries. They show that developed regions invest more in R&D

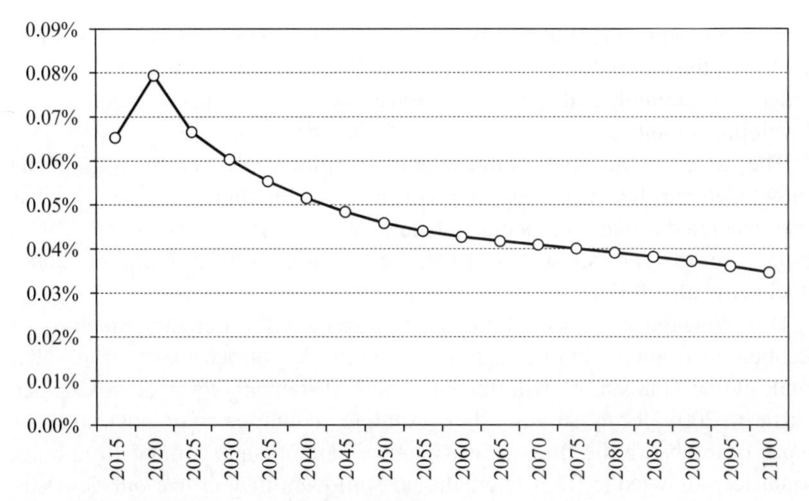

Figure 2.1 R&D as percentage of GWP

than in an efficient global policy scenario. International spillovers spread innovation and carbon-free backstop technologies globally. Without access to cheap abatement options in developing countries developed countries must invest more in knowledge creation at home.

Innovation is also very important if climate policy is efficient but some key mitigation technologies in the power sector – such as nuclear power or coal equipped with CCS – are not available. Many scenarios developed using WITCH show that it is optimal to react to technology constraints with higher innovation and diffusion of breakthrough technology in the power sector and with higher diffusion of wind and solar. Also in this case innovation starts first in developed regions, where both abatement costs and returns to innovation are higher. Innovation then spills over to non-OECD countries. The flexibility to invest in innovation greatly reduces the cost of climate policy when technology options are limited.

2.2 TECHNOLOGY COOPERATION: IMPLICATIONS OF GLOBAL R&D AND TECHNOLOGY DEPLOYMENT POLICIES

All emission scenarios generated by integrated assessment models show that the largest share of future incremental emissions will come from non-OECD countries. It is therefore of crucial importance that new technologies that increase energy efficiency and provide low- or zero-emission energy are quickly deployed in all developing countries and in particular in fast-growing emerging economies. It is thus not surprising that technology transfers are at the top of the international climate agenda. For example, the Bali Action Plan called for strengthened efforts to move technologies from developed to developing countries.

The idea of reducing atmospheric stabilisation costs by filling the knowledge gap between countries with more technological cooperation has also generated a lively academic debate since the early 1990s (Barrett, 1994, 2003; Carraro and Siniscalco, 1994; Grubb et al., 2002; Philibert, 2004; Buchner et al., 2005).

However, the enhanced circulation of ideas and the free dissemination of technological innovation throughout the world do not necessarily imply that total global innovation will increase and abatement costs decrease (see Carraro, 2001 for a survey). For example, countries may not have the capacity to absorb the flow of ideas and research results coming from other countries, knowledge from international spillovers may crowd-out domestic R&D efforts or free-riding incentives may induce some countries to reduce their own expenditures in R&D.

The existence of international R&D spillovers may justify international R&D policies to reduce the under provision of investment in new knowledge to combine with carbon pricing. For example an illustrative global fund can be considered, which is financed through a given share of each region's GDP, and provides a subsidy to each region that adds to their own expenditures on energy efficiency-improving R&D. This 'additionality' constraint is imposed because otherwise the R&D spending spurred by the subsidy would fully crowd out domestic R&D investments.[2] The size of the fund is equivalent to the global public R&D expenditures of the 1980s – or around 0.08 per cent of world GDP. In a scenario where such an R&D fund is established in association with a global cap-and-trade scheme designed to achieve a 550 ppm GHG concentration target, energy efficiency R&D increases significantly. However, compared with a carbon pricing (cap-and-trade) policy alone, the paths of energy intensity, carbon intensity of energy, the carbon price and GDP are only marginally affected. The same results are obtained when the effects of a global R&D policy are reassessed in the presence of breakthrough technologies. This is essentially because carbon pricing alone would already substantially boost energy efficiency and R&D in backstop technologies and because the international R&D spillovers to be internalised by the fund are limited. In this context, one open issue is whether a global fund to subsidise the deployment of already existing technologies may perform better. In the WITCH model, wind and solar power generation embody international learning spillovers, thus presenting potential gains from the internalisation of externalities. Following previous analysis, the fund is assumed to be equivalent to 0.08 percent of world GDP. Compared with carbon price alone, the additional investments in low carbon technologies alters significantly the energy mix. In particular, the technology deployment fund accelerates the diffusion of wind and solar electricity, whose share of world electricity production became much larger by 2050. However the impact on GDP costs remains negligible.

International spillovers may also be an important policy channel to increase equity and achieve higher efficiency. We can analyse, for example, the case of coupling a 550 ppm CO_2-eq stabilisation policy with a policy to foster knowledge dissemination in low income countries. The latter is implemented using a fraction of the revenues from emissions permit sales of developed countries to build absorption capacity in developing countries. The increased energy efficiency induced by the R&D cooperation policy generates a reduction of stabilisation costs in low income countries and, though costs in developed countries increase, a global reduction of GDP losses.

2.3 DIRECTED TECHNICAL CHANGE AND THE ROLE OF INTERSECTORAL SPILLOVERS

Virtually all climate-economy models find that climate change policies induce an increase in the pace of carbon-saving technical change. However, climate policy will also likely affect innovation not directly related to decarbonising the economy. A first point, raised by Nordhaus (2002), is that neglecting the competition between different forms of R&D would overestimate the benefit of endogenous technical change. Unfortunately, the majority of models used for climate policy analysis has only one R&D stock and imposes ad hoc assumptions to take into account the alternative and competitive uses of R&D funds, as, for example, Nordhaus (2002) and Popp (2004).[3]

The idea that traditional R&D efforts are crowded out by climate-related R&D investments is often cited in the modelling literature and originates from the hypothesis that the supply of R&D inputs is inelastic. This might indeed be a reasonable assumption in the short term because the number of scientists and laboratories is fixed and higher investments in research would simply shift researchers from one sector to another and raise their wages (Goolsbee, 1998). Nevertheless, in the long-run, the time horizon usually assumed to study climate policy, the allocation of total R&D investments across different sectors should not be constrained, because a strong stabilisation policy may induce higher expenditure in R&D to decarbonise the economy, while maintaining the same level of investment in other forms of R&D.

At the same time, it is not possible to rule out that forces other than the pure ‘crowding out’ effects shape the impact of climate policy on knowledge accumulation of other sectors. Goulder and Schneider (1999), Sue Wing (2003), Otto et al. (2007) and Gerlagh (2008) study the effect of climate policy on the equilibrium amount of R&D. These models share the hypothesis that technology advancements in both energy- and non-energy sectors are driven by a dedicated stock of knowledge. They all agree that climate policy modifies not only the direction of technical change, but also the total level of innovative activity. Goulder and Schneider (1999), Sue Wing (2003), and Otto et al. (2007) emphasise how the general equilibrium effect due to the policy-induced income reduction can lower the overall amount of resources available for knowledge creation. Gerlagh (2008) shows that if a sufficient amount of investments go to energy-saving technical change, then there might be a research dividend and overall research activity may increase.

Moreover, technical change is not neutral but it is fundamentally biased towards one input (Acemoglu, 2002). For example, capital-saving or labour-saving technical change might increase the demand of energy, i.e. they might be energy-biased. In this case it is important to understand whether the direction of innovation changes with relative prices, and how the distribution of investments in new knowledge creation changes as a consequence of climate policy.

To study the effects of climate policy on both the direction and the aggregate level of knowledge accumulation, the description of knowledge dynamics of the WITCH model has been enriched by a new module to endogenise technical change in the non-energy sector. R&D expenditures, and therefore knowledge accumulation, are factor-specific and can be directed towards increasing energy efficiency or towards rising productivity of non-energy inputs, namely capital and labour. In the model there are also mutual spillovers between the energy and non-energy sector (see www.witchmodel.org for a detailed description of the model, and Carraro et al., 2009 and Massetti and Nicita, 2010). The model abstracts from international spillovers, which, as shown in the previous section, have a modest role in shaping innovation dynamics.

By explicitly modelling two R&D capital stocks, it is possible to investigate whether there is any bias of technical progress and to study which policy induced forces, other than crowding-out, mould the reallocation of investments in knowledge of both energy and non-energy sectors. The inclusion of knowledge externalities between the two sectors also allows the study of the optimal portfolio of climate policies.

The Impact of Climate Policy on the Pace and the Bias of Technical Change

The first issue to be explored concerns the existence of a long-term bias of technical change. Table 2.1 summarises the BaU trends of R&D investments. The model features an upward sloping path of non-energy R&D as share of GWP and a path of energy R&D as share of GWP slightly declining. The relative share of energy R&D over total R&D is also declining from about one percent to 0.71 percent. As a result, total R&D expenditure is increasing and driven by non-energy R&D investments.

Table 2.1 Baseline trends of R&D

	2005	2025	2045	2065	2085	2100
R&D expenditure (%GWP)	2.15	2.24	2.30	2.38	2.45	2.46
Non-energy R&D (%GWP)	2.13	2.22	2.28	2.36	2.43	2.44
Energy efficiency R&D (%GWP)	0.0216	0.0189	0.0181	0.0240	0.0178	0.0174
Energy R&D (% total investment in R&D)	1.01	0.84	0.79	0.76	0.73	0.71

Technical change is therefore mainly capital-labour augmenting and this trend is even reinforced across the century because wages, endogenously determined, increase faster than equilibrium fuel prices.

Most interestingly, capital-labour augmenting technical change is energy-biased because the capital-labour aggregate and energy are complements, as found by empirical studies and also assumed in the model. This means that R&D investments directed towards the capital-labour aggregate increase energy use. This outcome of the model implies that capital-labor-augmenting technical change is carbon-biased and knowledge advancements *per se* are not necessarily good for the environment.

The stabilisation policy has a remarkable impact on R&D dynamics as the comparison between Table 2.1 and Table 2.2 clearly shows. It induces a substantial increase of energy R&D investments to improve energy efficiency, and lowers non-energy R&D investments with respect to the BaU scenario.

The need to increase energy efficiency modifies the equilibrium ratio of energy R&D to non-energy R&D investments from a declining path, over the century, to a rising one. Therefore, in an unconstrained economy (without a stabilisation target), R&D is directed towards augmenting the productivity of the capital-labour aggregate, which becomes relatively scarcer with respect to energy, as population growth slows down worldwide. In a carbon constrained economy, instead, the price of the energy input grows faster than the price of

Table 2.2 Stabilisation trends of R&D

	2005	2025	2045	2065	2085	2100
R&D expenditure (%GWP)	2.12	2.21	2.19	2.25	2.31	2.32
Non-energy R&D (%GWP)	2.09	2.14	2.13	2.19	2.25	2.27
Energy efficiency R&D (%GWP)	0.0265	0.0304	0.0390	0.0740	0.0382	0.0356
Energy R&D (%Total Investment in R&D)	1.25	1.38	1.78	1.80	1.65	1.54

the non-energy input, and thus R&D resources are directed to increase energy efficiency.

Nonetheless, the drop in non-energy R&D is larger than the sharp increase in energy R&D. As a consequence, at the equilibrium, overall expenditure in R&D decreases. This result confirms the findings of Goulder and Schneider (1999) and Sue Wing (2003) and Otto et al. (2007).

What are the economic forces behind this result? Does this contraction result from a crowding-out effect of increased investment in energy R&D?

The results of the analysis, as shown in detail by Carraro et al. (2009), find that there is no direct competition between the energy and non-energy R&D and the reduction of the rate of investments in non-energy R&D is not determined by a pure crowding-out effect. Quite the opposite, the two stocks of knowledge display a mild degree of complementarity. Reduced spending in non-energy R&D is instead due to: 1) a general contraction of economic activity induced by the stabilisation policy, which in turn induces a lower demand of capital-labour services and a lower investment in non-energy R&D; 2) the fact that non-energy augmenting technical change is energy biased because of the complementarity between the energy and the non-energy sector. With energy biased technical change, an increase of non-energy R&D spending would increase energy use, and vice versa: by reducing non-energy R&D spending it is possible to reduce energy demand, an important way to cut emissions in a stabilisation scenario. It is therefore the stabilisation policy itself that induces a contraction of the optimal level of R&D in the non-energy sector, and not the competition from higher spending in energy R&D. Carraro et al. (2009) widely discussed this result and argued against the exogenous crowding-out hypothesis imposed in Nordhaus (2002) and Popp (2004, 2006) on the grounds that, at least in the medium/long-term, societies are free to allocate the optimal amount of resources to knowledge creation. Recent empirical evidence presented in Popp and Newell (2009) confirms this intuition, showing that increased spending in energy R&D does not necessarily crowd-out non-energy R&D.

The Optimal Climate Portfolio when Knowledge Spills across Sectors

Because of knowledge externalities, even in the presence of a stabilisation policy, global R&D spending may remain below levels that would be socially optimal. Climate policy may even worsen the gap between the actual and the socially optimal level of investments in innovation by weakening the incentives to invest in knowledge accumulation outside the energy sector.

Many researchers that have worked on the optimal design of climate policy have stressed the importance of studying climate policy in a second-best setting with both environmental and knowledge externalities. For example, Jaffe et al. (2005) proposes to use a portfolio made of a price signal to correct for the environmental externality coupled with a policy to support investment in technologies to reduce GHG emissions. The idea of complementing a stabilisation policy with an R&D policy in order to address both externalities at once is instead opposed by Nordhaus (2011). He argues that once the environmental externality is corrected, there are no evident reasons to treat research in technologies that reduce GHG emissions, differently from other kinds of research that share the same characteristic of the public good.

To shed some light on the questions raised by the ongoing debate, it is helpful to assess whether the stabilisation policy contributes to bringing the economy closer to or farther away from the socially optimal level of innovation, and the welfare implications of implementing a policy mix that combines a stabilisation policy and R&D policies to support the optimal level of innovation.

As for the first issue, by comparing the optimal path of R&D investments with the path characterising both the BaU scenario and the stabilisation scenario,[4] we are in a good position to evaluate how effective a policy is that addresses the only environmental externality at closing the gap between the level of innovation of both the energy and the non-energy sector and the socially optimal one. Figure 2.2 and Figure 2.3 show the time path of R&D investments – as percentage of gross world product (GWP) – when the stabilisation policy is implemented and domestic knowledge externalities are internalised. The optimal path of investments in energy efficiency R&D is characterised by a declining trend over the century. The opposite is true for the optimal time path of non-energy R&D investments: the trend is increasing because labour becomes a scarce resource as population growth levels off by mid-century. The difference between the optimal path and the second-best scenarios is striking. If we consider energy R&D, the stabilisation policy brings R&D investments closer to the socially optimal level. Remarkably, the jump from the level optimal in the BaU does not close the R&D gap. Contrary to what happens in energy R&D, the stabilisation

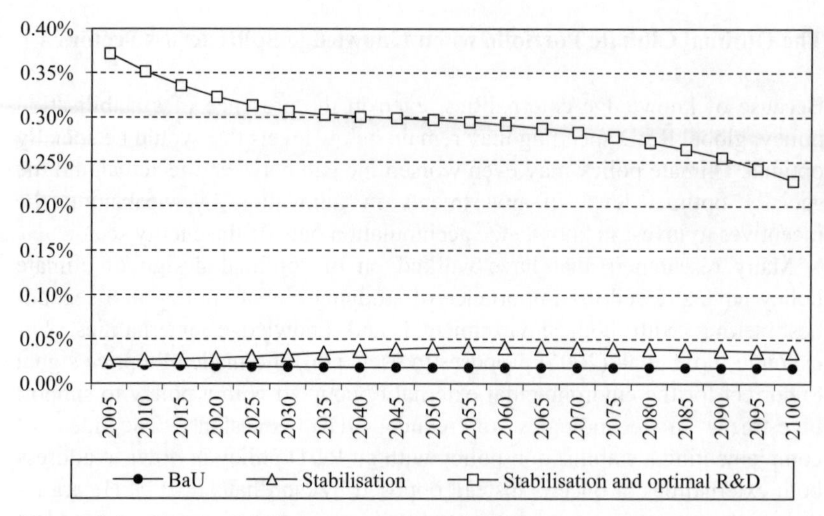

Figure 2.2 Energy R&D as percentage of GWP

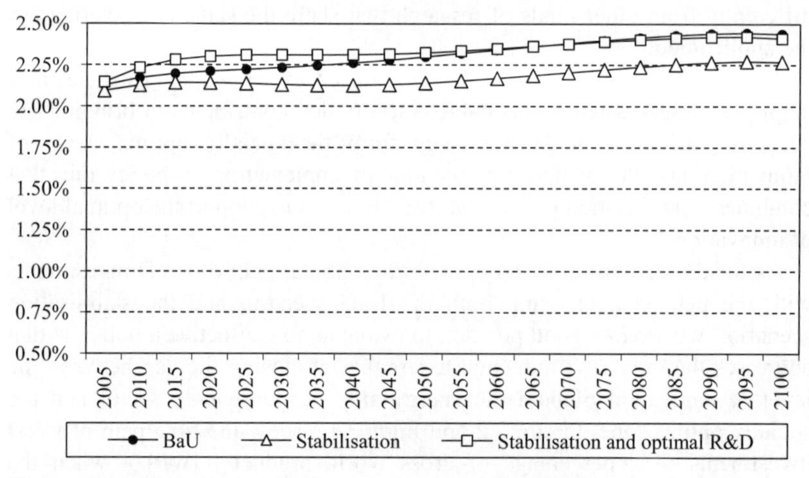

Figure 2.3 Non-energy R&D as percentage of GWP

policy brings investments in non-energy R&D down. Consequently, total R&D investment moves farther away from the optimal level.

We have so far shown that only addressing the environmental externality is, at best, not sufficient to bring R&D investments to the socially desirable level. In fact, the environmental policy exacerbates the knowledge externality in the non-energy sector. Therefore, at least in our modelling context, policies that address both externalities appear to be socially desirable.

Figure 2.4 assesses the impact of R&D policy when coupled with the stabilisation policy and shows that policy addressed at internalising knowledge externalities in the energy sector has a remarkable impact on stabilisation costs: combining an energy R&D policy to the stabilisation policy would reduce costs to 0.14 percent of GDP for OECD countries and would also cut them considerably in non-OECD ones. At the global level, stabilisation costs would be reduced to roughly one fourth of what they would be without the energy R&D policy. As expected, the energy R&D policy has a greater impact on costs in OECD countries, were the bulk of the knowledge externality is found. Figure 2.4 also shows that internalising all knowledge externalities reduces stabilisation costs further, even if by a lesser extent than the energy R&D policy. Stabilisation costs virtually disappear for OECD countries. For non-OECD countries the reduction of costs is less pronounced, as expected, and at global level internalising non-energy R&D externalities reduces stabilisation costs of 0.1 percent of discounted GWP.

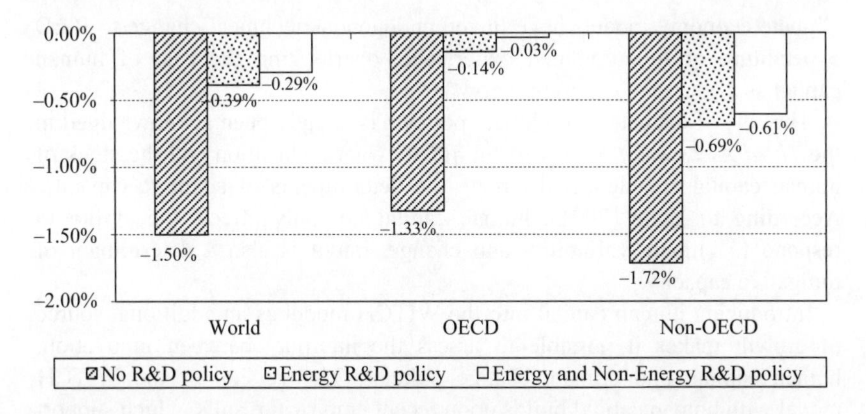

Figure 2.4 Discounted stabilisation policy costs as percentage of GWP

The result that complementing the stabilisation policy with an R&D policy brings a reduction of stabilisation costs is in line with the findings of the two main studies, Goulder and Schneider (1999) and Popp (2006). These studies analyse a climate policy portfolio in which R&D policy is coupled with a policy to reduce GHG emissions by means of computational models with knowledge externalities. However, there are some important differences between the three models and the policies examined. Goulder and Schneider (1999) focus on intrasectoral spillovers and find that an R&D policy reduces stabilisation costs only if it addresses R&D externalities in all sectors. If restricted only to sectors with low emissions, the R&D policy increases stabilisation costs. Popp (2006) shows that higher spending in energy R&D reduces stabilisation costs only marginally, because it crowds-out non-energy

R&D investments. The crowding-out is exogenous because Popp does not model the explicit knowledge accumulation in the non-energy sector. Contrary to Popp (2006) we do not impose exogenous crowding-out assumptions because we model both knowledge stocks. We find that a stabilisation policy along with an R&D policy targeted at the only energy sector is significantly less costly than the stabilisation policy alone. We find that energy R&D does not crowd-out non-energy R&D and, thanks to intersectoral spillovers, the policy induced increase in energy efficiency R&D spills over to the non-energy sector, contributing to knowledge accumulation and the reduction of knowledge externalities.

2.4 HUMAN CAPITAL FORMATION AND MITIGATION POLICY

Climate economy models have linked endogenous technical change to R&D expenditure or to cumulated experience, overlooking the role of human capital as a source of economic growth.

The role of education in climate policy has already been acknowledged in the *Third Assessment Report* of the IPCC, where education and the stock of human capital are identified among the determinants of adaptive capacity. According to Yohe (2001), human capital not only affects the ability to respond to climate variability and change, but it is also a determinant of mitigative capacity.

Introducing human capital into the WITCH model as an additional source of growth makes it possible to assess the interplay between innovation, human capital, and climate change policies. The version of the WITCH model with human capital builds upon recent empirical results, which support the hypothesis that the technology drivers of factor productivities are input-specific (Carraro et al., 2012). Whereas human capital is an important driver of labour productivity, innovation, measured by the accumulated stock of knowledge, enhances capital and energy productivity (see www.witchmodel.org and Carraro et al., 2012 for a detailed description of the model).

Starting from the basic set-up described in Section 2.1, the macroeconomic production structure is modified. The three factors of production, labour, energy and capital, are modelled as gross complements, as most econometric studies suggest, and factor-augmenting technical change with dedicated technology drivers, as estimated in Carraro and De Cian (2009) is introduced. In the model, capital productivity is enhanced by the stock of generic knowledge, energy productivity is raised by both the stock of generic knowledge and the stock of energy knowledge while labour productivity depends on human capital.

The production of both human capital and knowledge is characterised by intertemporal spillovers. The stock available in each region at a given point in time contributes to the creation of the future stock. Following state-of-the-art literature (Romer 1990, Jones 1995, Popp 2002, Glomm and Ravikumar 1992, Blankenau and Simpson 2004) human capital is produced using a Cobb-Douglas combination of the existing stock of human capital and current expenditure in education. In a similar way, the available knowledge stock and current R&D investments are combined to produce new knowledge. The creation of energy knowledge is also influenced by international spillovers.[5]

For a correct interpretation of the results it is important to stress that, with gross complementarity between factors of production and positive elasticity of human capital on labour productivity, human capital has an energy-using effect. As for R&D, the direct impact on energy demand is negative (e.g. energy-saving) if the elasticity of substitution is less than one. However, the indirect impact via capital productivity is energy-using, as in the case of human capital. The net effect ultimately depends on the relative size of the elasticity of capital productivity and energy productivity with respect to knowledge. Carraro and De Cian (2009) find that that the former is lower than the latter suggesting that overall generic R&D has an energy-saving effect. The implications of a climate policy for the energy sector and energy-saving innovation are similar to those obtained in the basic set up of the model described in Section 2.1 and the model with directed technical described in Section 2.3.

When facing a climate policy constraint, each region reshapes the optimal mix of investments to meet the constraint at the minimum cost. The carbon price signal reallocates resources towards low carbon technologies (renewable energy, coal equipped with carbon capture and storage, and nuclear), energy efficiency R&D, clean energy R&D, and subsequently to the deployment of the breakthrough technologies.

Results reported in Figure 2.5 show that climate policy stimulates investments in both energy and general purpose innovation. This is because general purpose R&D, as mentioned above, is overall energy-saving and therefore a cost-effective abatement option.

This result differs from previous findings that considered different R&D programs (Goulder and Schneider 1999, Sue Wing 2003, and the model described in Section 2.3). For example, the WITCH model with directed technical change finds that because non-energy R&D is energy-using, climate policy reduces R&D investments of the entire economy.

In percentage terms, climate policy reallocates relatively more resources to energy R&D because it is more effective at augmenting energy efficiency than generic R&D.

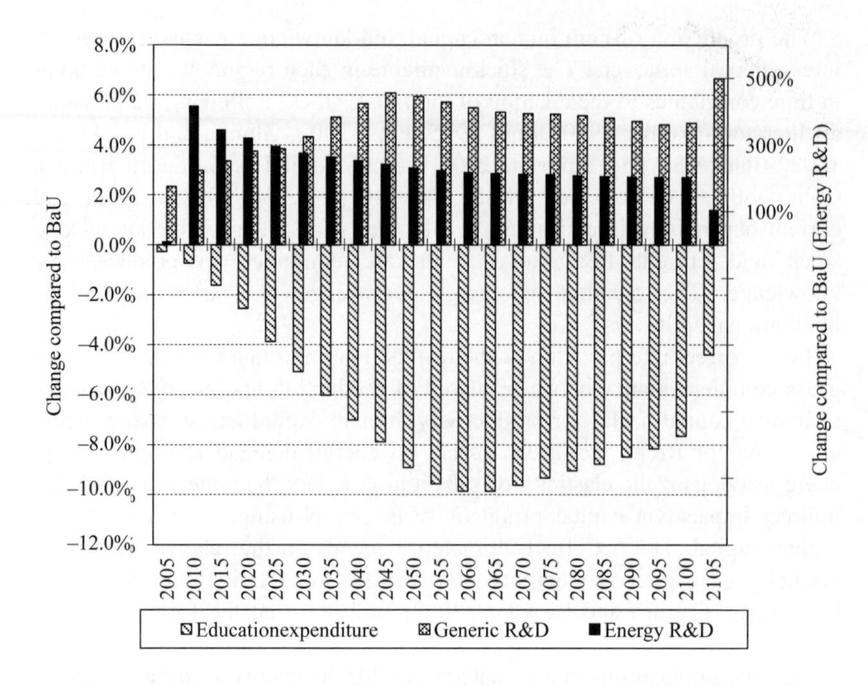

Figure 2.5 Investments in education, generic R&D and energy R&D. Percentage change with respect to BaU

The result that climate policy stimulates both investments in energy and generic R&D is in line with recent empirical evidence, which confirmed that crowding-out between clean and dirty energy R&D might exist, but not between innovation in the energy and non-energy sector (Popp and Newell, 2009).

As to the impact of climate policy on education expenditure, results, as shown in Figure 2.5, are in line with the findings of the previous section. Since education is energy-using, climate policy reallocates productive resources away from human capital formation and towards R&D, crowding-out education expenditure by at most 10 percent at mid-century. As already anticipated, this is due to the fact that human capital is labour-augmenting and to the complementarity between energy and labour. This is a result that has been already found in the literature on environmental policy and human capital. When pollution is linked to final output, environmental policy can reduce education expenditure and slow down human capital accumulation (Gradus and Smulders, 1993; Hettich, 1998; Pautrel, 2012).

This result points at the potential trade-off with other policy goals government might have. For example, universal primary education and sustainable development are two of the eight Millennium Development Goals

(MDG) that 189 countries have committed to achieve by 2015. The *Fourth Assessment Report* itself (IPCC, 2007) has stressed the importance of capacity building and socio-economic development, both dependent upon education, for effective climate policy.

Against this background, a combination of climate and education policy is studied. The education policy requires Sub-Saharan Africa and South Asia (SSA and SASIA) to increase education investments so that the fraction of population currently off-track is on-track from 2015 onwards.[6] The remaining regions are required to maintain the path of education expenditure foreseen in the no-climate policy case, as current spending is already consistent with the achievement of the MDG.

The macroeconomic effects of combining education and climate policy are shown in Table 2.3.

Table 2.3 Macroeconomic effects of combining education and climate policy

NPV	Gross world product	Consumption	Education	Generic R&D	Energy R&D
Climate and education policy	−1.03%	−1.12%	1.13%	4.60%	316.14%
Climate policy	−1.37%	−1.09%	−5.31%	4.15%	317.97%

Adding the education policy stimulates investments in generic R&D, which has a direct impact on factor productivities and thus on economic growth. The increase in education expenditure puts an upward pressure on emissions as well. However, the impact on the carbon market and energy mix is very moderate. The carbon price increases, but only slightly (at most by 2 percent at the end of the century) and the effect on output growth partially compensates the costs of climate policy. In Net Present Value, climate mitigation costs are lower in terms of gross world output, but higher in terms of consumption. This result raises the issue of the appropriate metric to measure the costs of a policy (Hourcade and Ghersi, 2008). Whereas output provides a measure of the macroeconomic effects, consumption is a better indicator of welfare. In addition, net present values are aggregate figures that hide a trade-off between short-term and long-term consumption. Figure 2.6 shows that in the short-term, education policy absorbs additional resources, reducing consumption possibilities. However, additional education expenditure pays off in the long-term, when it increases overall economic growth, and ultimately consumption.

The model just described only accounts for the direct effect of human capital on labour productivity. It neglects any indirect effect human capital

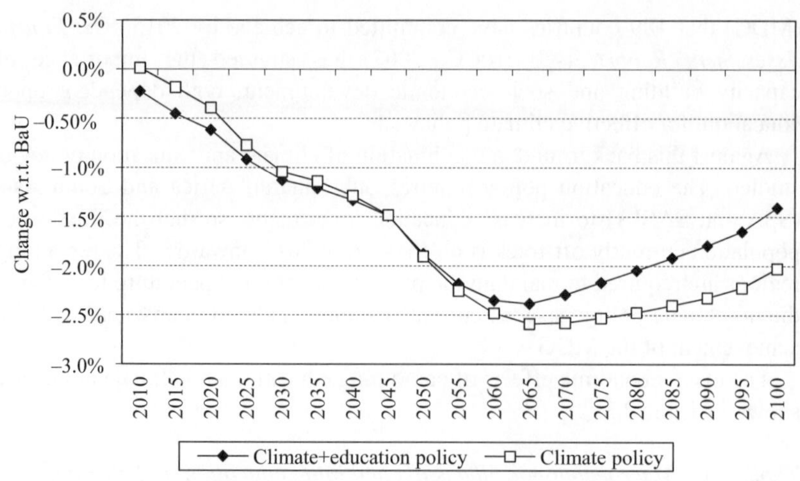

Figure 2.6 Consumption path in the policy scenarios. Percentage change with respect to BaU

could have on the innovation process, by facilitating technology absorption and new knowledge creation, The main reason behind this modelling choice is the lack of empirical guidance on how to model the indirect link. An explorative exercise is illustrated to show the implications of different formulations. A first attempt broaden the role of human capital that now contribute at increasing the ability to adopt new technologies. This idea is integrated into the model by assuming that the absorptive capacity improves not only by the energy knowledge stock, but also by building up human capital. In the second model variant human capital is an essential input in the creation of both stocks of generic and energy knowledge. Table 2.4 shows that, although modelling the effects of human capital on absorption capacity and knowledge formation is able to lessen the crowding-out induced by climate policy, energy-using effect is shown to prevail.

Table 2.4 Impact of climate policy on education expenditure when considering alternative model specifications

Basic model	Basic model + effect on absorptive capacity	Basic model + effect on knowledge creation
–8.4% (–76.7)	–8.0% (–74.3)	–5.3% (–54.3)

Note: Percentage change of cumulative investments compared to BaU (USD2005 Trillion in parenthesis).

2.4 CONCLUSIONS

This chapter has analysed the interplay between technical change and climate policy with a twofold goal: first to investigate the effects of climate policy on the investments in new knowledge and technology deployment and second to assess the impact of R&D policy or education policy on the cost of climate policy.

Results of scenario analysis with the WITCH model reveals that a carbon price has sizeable effects on R&D and technology deployment. However, to achieve major technological breakthroughs, a strong price signal is needed to spur the necessary investments.

When the potential role of international knowledge flows in fostering the development of new energy technologies is assessed, results suggest that a global R&D fund to subsidise R&D and/or low-carbon technology deployment could reduce mitigation costs if it came on top of a carbon price. However, the optimal size of such a fund, and its effects, are found to be typically small.

When directed technical change and intersectoral spillovers are explicitly modelled, results reveal that climate policy internalises only partially knowledge externalities in the energy sector and it even worsens market failures in the non-energy sector. Correcting the environmental externality alone has contrasting effects on the knowledge externality. Given the relative size of the two sectors, the stabilisation policy induces a lower amount of R&D spending than in the BaU. Thus the stabilisation policy brings us farther from the optimal level of R&D spending, increasing the need for policies to correct for the knowledge externality, which has been shown to reduce substantially climate policy costs.

The different modelling approaches to endogenous technical change proposed in Section 2.3 and 2.4 have shown how climate policy can have different impacts on innovation and human capital formation. It is clear that when energy on one hand and labour and capital on the other hand are complements – as suggested by many empirical studies – any capital-labour augmenting technical change is fundamentally a pollution-biased technical change and is thus discouraged by price signals induced by climate policy. Although a stabilisation policy provides stimulus for energy saving innovation, the overall effect can be a reduction of the knowledge and human capital intensity of the economy. These results should be taken with caution because indirect effects that could be energy-saving are not captured. For example, the relationship between human capital and innovation is not explicitly accounted for. When included, these indirect channels would mitigate the contraction effect induced by climate policy.

The consequence for studies that aim at assessing the economic cost of climate policy is as follows. The role of technical change to achieve GHG concentration stabilisation targets is crucial. If a study neglects the impact of climate stabilisation on energy saving technical change, it overestimates the cost of climate policy. However, if a study considers the impact of climate policy only on energy-related and carbon-free technical change, the cost of climate policy will be underestimated. A proper assessment of the cost of climate policy must take into account the contrasting effects induced on both energy and non-energy R&D and related consequences.

NOTES

1. The present chapter is based on Bosetti et al. (2008), Bosetti et al. (2009b), Carraro et al. (2009), Massetti and Nicita (2010), Carraro et al. (2012).
2. To implement the additionality constraint, a minimum level of energy efficiency R&D investment is imposed in all regions, equal to what would be Pareto optimal under a 550 ppm GHG concentration stabilisation scenario.
3. Also the basic version of WITCH used for most of the studies presented in this book describes innovation dynamics only in the energy sector.
4. Here we define an optimal world as one in which the stabilisation policy is implemented to correct the environmental externality and knowledge intersectoral externalities that are fully internalised in each region. In the Business-as-Usual scenario no externalities are internalised, In the stabilisation scenario only the environmental externalities is addressed.
5. Although it would be natural to characterise spillovers in the general purpose R&D sector, we refrain from doing so, mostly because of consistency with the empirical study that is used to calibrate our model, which did not account for spillovers. In addition, previous studies (see Bosetti et al., 2008) show that the contribution of knowledge spillovers is limited.
6. Countries or population are classified on-track in achieving universal primary education if continuing on linear trends between 1990 and 2002 will result in a completion rate above 95 percent by 2015. Off- track means that the completion rate is projected to be below 50 percent in 2015 (seriously off track) or below 95 percent (moderately off track).

REFERENCES

Acemoglu, D. (2002), 'Directed technical change', *Review of Economic Studies*, **69**(4), 781–809.

Barrett, S. (1994), 'Self-enforcing international environmental agreements', *Oxford Economic Papers*, special issue on 'Environmental economics', **46**(3), 878–894.

Barrett, S. (ed.) (2003), *Environment and Statecraft: The Strategy of Environmental Treaty-Making*, Oxford: Oxford University Press.

Blankenau, W. and N. Simpson (2004), 'Public education expenditure and growth', *Journal of Development Economics*, **73**(2), 583–605.

Bosetti, V., C. Carraro, M. Galeotti, E. Massetti and M. Tavoni (2006), 'WITCH: A World Induced Technical Change Hybrid Model', *The Energy Journal*, special issue on 'Hybrid modeling of energy- environment policies: Reconciling bottom-up and top-down', **27**(SI2), 13–38.

Bosetti, V., C. Carraro, E. Massetti, A. Sgobbi and M. Tavoni (2009a), 'Optimal energy investment and R&D strategies to stabilise greenhouse gas atmospheric concentrations', *Resource and Energy Economics*, **31**(2), 123–137.

Bosetti, V., C. Carraro, R. Duval, A. Sgobbi and M. Tavoni (2009b), 'The role of

R&D and technology diffusion in climate change mitigation: New perspectives using the WITCH model', OECD Economics Department Working Paper No. 664.
Bosetti, V., C. Carraro, E. Massetti and M. Tavoni (2008), 'International energy R&D spillovers and the economics of greenhouse gas atmospheric stabilization', *Energy Economics*, **30**(6), 2912–2929.
Bosetti, V., E. Massetti and M. Tavoni (2007), 'The WITCH model: Structure, baseline and solutions', Nota di Lavoro 10.2007, Milan: Fondazione Eni Enrico Mattei.
Buchner, B., C. Carraro, I. Cersosimo and C. Marchiori (2005), 'Back to Kyoto? US participation and the linkage between R&D and climate cooperation', in A. Haurie and L. Viguier (eds), *The Coupling Climate and Economic Dynamics*, Dordrecht: Kluwer Academic Publishers, pp. 173–204.
Carraro, C. (2001) 'Environmental technological innovation and diffusion', in H. Folmer, L. Gabel, S. Gerking and A. Rose (eds), *Frontiers of Environmental Economics*, Cheltenham, UK and Northampton, MA, USA: Edward Elgar, pp. 342–370.
Carraro, C. and E. De Cian (2009), 'Factor-augmenting technical change: An empirical assessment', Nota di Lavoro 18.2009, Milan: Fondazione Eni Enrico Mattei.
Carraro, C., E. De Cian, L. Nicita, E. Massetti and E. Verdolini (2010), 'Environmental policy and technical change: A survey', *International Review of Environmental and Resource Economics*, **4**(2), 163–157.
Carraro, C., E. De Cian and M. Tavoni (2012), 'Human capital, innovation and climate policy: An integrated assessment', Nota di Lavoro 18.2012, Milan: Fondazione Eni Enrico Mattei.
Carraro, C., E. Massetti and L. Nicita (2009), 'How does climate policy affect technical change? An analysis of the direction and pace of technical progress in a climate-economy model', *The Energy Journal*, special issue on 'Climate change policies after 2012', **30**(2), 7–38.
Carraro, C. and D. Siniscalco (1994), 'Environmental policy re-considered: The role of technological innovation', *European Economic Review*, **38**(3–4), 545–554.
De Cian, E., V. Bosetti and M. Tavoni (2011), 'Technology innovation and diffusion in less than ideal climate policies. An assessment with the WITCH model', *Climatic Change*, **114**(1), 121–143, DOI: 10.1007/s10584-011-0320-5.
Gerlagh, R. (2008), 'A climate-change policy induced shift from innovations in carbon-energy production to carbon-energy savings', *Energy Economics*, **30**(2), 425–448.
Gillingham, K., R.G. Newell and W.A. Pizer (2008), 'Modeling endogenous technological change for climate policy analysis', *Energy Economics*, **30**(6), 2734–2753.
Glomm, G. and B. Ravikumar (1992), 'Public versus private investment in human capital: Endogenous growth and income inequality', *Journal of Political Economy*, **100**(4), 818–834.
Goolsbee, A. (1998), 'Does government R&D policy mainly benefit scientists and engineers?', *The American Economic Review*, **88**(2), 298–302.
Goulder, L.H. and S.H. Schneider (1999), 'Induced technological change and the attractiveness of CO_2 abatement policies', *Resource and Energy Economics*, **21**(3–4), 211–253.
Gradus, R. and S. Smulders (1993), 'The trade-off between environmental care and long-term growth-pollution in three prototype growth models', *Journal of Economics*, **58**(1), 25–51.
Griliches, Z. (1957), 'Hybrid corn: An exploration in the economics of technological change', *Econometrica*, **25**(4), 501–522.

Griliches, Z. (1992), 'The search for R&D spillovers', *Scandinavian Journal of Economics*, **94**(0), 29–47.
Grossman, G.M. and E. Helpman (1991), *Innovation and Growth in the Global Economy*, Cambridge, MA: MIT Press.
Grubb, M., C. Hope and R. Fouquet (2002), 'Climatic implications of the Kyoto protocol: The contribution of international spillover', *Climatic Change*, **54**(1–2), 11–28.
Hall, B. (1996), 'The private and social returns to research and development', in B. Smith and C. Barfield (eds), *Technology, R&D, and the Economy*, Washington, DC: Brookings, pp. 140–183.
Hettich, F. (1998), 'Growth effects of a revenue-neutral environmental tax reform', *Journal of Economics*, **67**(3), 287–316.
Hourcade, J.C. and F. Ghersi (2008), 'Interpreting environmental policy cost measures', in V. Bosetti, R. Gerlagh and S. Schleicher (eds), *Modeling Transitions to Sustainable Development*, Cheltenham, UK and Northampton, MA, USA: Edward Elgar, pp. 61–82.
IPCC (Intergovernmental Panel on Climate Change) (2007), *Climate Change 2007: Mitigation, Contribution of Working Group III to the Fourth Assessment Report of the Intergovernmental Panel on Climate Change* [B. Metz, O.R. Davidson, P.R. Bosch, R. Dave, L.A. Meyer (eds)], Cambridge, United Kingdom and New York, NY, USA: Cambridge University Press.
Jaffe, A.B. (1986), 'Technological opportunity and spillover of R&D: Evidence from firms patents, profits, and market value', *American Economic Review*, **76**(5), 984–1001.
Jaffe, A.B., R.G. Newell and R.N. Stavins (2005), 'A tale of two market failures: Technology and environmental policy', *Ecological Economics*', **54**, 164–174.
Jones, C. (1995), 'R&D based models of economic growth', *Journal of Political Economy*, **103**(4), 759–784.
Jones, C.I. (1999), 'Growth: With or without scale effects?', *American Economic Review*, **89**(2), 139–144.
Jones, C.I. and J.C. Williams (1998), 'Measuring the social return to R&D', *The Quarterly Journal of Economics*, **113**(4), 1119–1135.
Kouvaritakis, N., A. Soria and S. Isoard (2000), 'Endogenous learning in world post-Kyoto scenarios: Application of the POLES model under adaptive expectations', *International Journal of Global Energy Issues*, **14**(1–4), 222–248.
Li, C.W. (2000), 'Endogenous vs. semi-endogenous growth in a two-R&D-sector model', *The Economic Journal*, **110**(14), 109–122.
Mansfield, E. (1977), 'Social and private rates of return from industrial innovations', *Quarterly Journal of Economics*, **91**(2), 221–240.
Mansfield, E. (1996), 'Microeconomic policy and technological change', in J.C. Fuhrer and J.S. Little (eds), *Technology and Growth: Conference Proceedings*, Boston, MA: Federal Reserve Bank of Boston, pp. 183–200.
Massetti E. and L. Nicita (2010), 'Optimal R&D investments and the cost of GHG stabilization when knowledge spills across sectors', CESifo Working Papers No. 2988.
Nemet, G.F. (2006), 'Beyond the learning Ccurve: Factors influencing cost reductions in photovoltaics', *Energy Policy*, **34**(17), 3218–3232.
Nordhaus, W.D. (1991), 'Economic approaches to greenhouse warming', in R. Dornbusch and J.M. Poterba (eds), *Global Warming: Economic Policy Responses*, Cambridge, MA: MIT Press, pp. 33–68.

Nordhaus, W.D. (2002), ‘Modeling induced innovation in climate change policy’, in A. Grubler, N. Nakicenovic and W.D. Nordhaus (eds), *Modeling Induced Innovation in Climate Change Policy*, Washington, DC: Resources for the Future Press, pp. 275–314.

Nordhaus, W.D. (2011), ‘Designing a friendly space for technological change to slow global warming’, *Energy Economics*, **33**(4), 665–673.

Otto, V.M., A. Loschel and R. Dellink (2007), ‘Energy biased technical change: A CGE analysis’, *Resource and Energy Economics*, **29**(2), 137–158.

Pautrel, X. (2012), ‘Environmental policy, education and growth: A reappraisal when lifetime is finite’, *Macroeconomic Dynamics*, **16**(5), 661–685.

Philibert, C. (2004), ‘International energy technology collaboration and climate change mitigation’, Paris: OECD Environment Directorate and International Energy Agency.

Popp, D. (2002), ‘Induced innovation and energy prices’, *American Economic Review*, **92**(1), 160–180.

Popp, D. (2004), ‘ENTICE: Endogenous technological change in the DICE model of global warming’, *Journal of Environmental Economics and Management*, **48**(1), 742–768.

Popp, D. (2006), ‘R&D subsidies and climate policy: Is there a “free lunch”?’, *Climatic Change*, **77**(3–4), 311–341.

Popp, D. and R.G. Newell (2009), ‘Where does energy R&D come from? Examining crowding out from environmentally-friendly R&D’, NBER Working Paper No. 15423.

Romer, P. (1990), ‘Endogenous technical change’, *Journal of Political Economy*, **98**(5), 71–102.

Sue Wing, I. (2003), ‘Induced technical change and the cost of climate policy’, MIT Joint Program on the Science and Policy of Global Change, Report No. 102.

Yohe, G. (2001), ‘Mitigative capacity: The mirror image of adaptive capacity on the emissions side’, *Climatic Change*, **49**(3), 247–262.

3. Getting to Yes

Alessandra Sgobbi

3.1 INTRODUCTION

Global warming has emerged as the defining challenge of our century, and prominently features in international and national political agendas. Interestingly, it is no longer confined to the environmental realm, but rather it is discussed in high-level economic forums, such as Davos and the G8 meetings, where the link with economic growth and prosperity is made explicit.[1]

Climate change is truly a global problem, and addressing it requires unprecedented levels of collaboration and coordinated actions. Yet, a cursory look at the literature and at recent news shows that, despite alleged progress at the latest Conferences of the Parties of the United Nations Framework Convention on Climate Change (UNFCCC) in Cancun (2010), Durban (2011) and Doha (2012), we are still far from achieving a consensus on the steps and measures needed to fully address the challenge. For the very first time in the history of the UNFCCC, the so-called 'Cancun Agreements' bring the global goal of keeping the temperature increase below 2°C into formally negotiated text.[2] However, in Cancun, delegates failed to agree on concrete mitigation measures and, above all, on how the burden should be shared across countries and regions. The 17th Conference of the Parties in Durban reaffirmed and furthered the commitments taken in Cancun the previous year, in particular on climate finance. Yet, parties failed once again to agree on concrete mitigation targets. The 'Doha Gateway' does pave the way for the continuation of the Kyoto Protocol, yet the level of ambition is well below what would be required to keep temperature increase below 2°C, with major emitting countries remaining outside the second commitment period of the Kyoto Protocol.

There are indeed many political and historical reasons for this impasse in international efforts. Countries have yet to come to a shared understanding of the implications of the Rio principles of common but differentiated responsibility and respective capabilities. While an 'equitable' deal should abide to these principles, any deal will have limited environmental effectiveness without firm commitments from both developed and developing countries, particularly the more advanced and fast growing economies. Developing countries are reluctant to commit to action unless a clear proposal is made on the financial resources that will be available to them for mitigation and adaptation. Substantial disparities exist in welfare levels, and the right to development cannot be compromised, as poverty eradication is still the overarching priority.

Despite these historical, political, and ideological differences, it is the economic implications of an international agreement on climate change that will, to a large extent, determine its acceptability and feasibility.

An environmentally successful international climate policy framework needs to bring the main greenhouse gases (GHG) emitting countries together into a climate coalition that delivers ambitious emission reductions. For such an agreement to also be economically efficient, it would need to enable its implementation through the least costs policies and instruments.

First, an international climate agreement needs to be both wide and stable. Broad based country participation is required for any agreement to be environmentally effective. This is particularly relevant if we consider that contributions to global GHG emissions are dynamic – that is, current main emitters may not be the major emitters in the future, if economic growth and development continue along the current path. Yet, wide coalitions may be harder to achieve and/or maintain, as individual countries have incentives to free-ride on the efforts of others, benefitting from emission reduction while not paying for it.

Second, the timing of reaching a global agreement has serious implications on the costs of achieving the targets, and thus on their feasibility. Robust decision-making under uncertainty requires that policy makers consider the possibility of such an agreement in the future when making decisions today.

Third, the timing of participation of different countries in mitigation efforts is essential in determining the policy outcome and its burden sharing. While today most Non-Annex I (NA1) countries firmly reject a binding cap on their emissions, their participation is essential if we want to have a reasonable chance to keep the increase of mean global temperature below 2°C.

This chapter explores the implications of the following three problems: (i) breadth and depth of different configurations of global actions (Section 3.2) (ii) uncertainty over the timing of an agreement and the timing of participation of different regions of the world (Sections 3.3 and 3.4), and (iii) the implications of different types of international agreements in terms of environmental effectiveness, economic efficiency, and political feasibility (Section 3.5). The results will shed some light on the key elements that are needed to reach a realistic agreement in dealing with climate change. The analyses are performed using the WITCH model.

WITCH (Bosetti et al., 2006) is a regional integrated assessment hard-link hybrid climate-energy-economy model designed to assist in the study of the socio-economic dimension of climate change. It is structured to provide information on the optimal responses of world economies to climate damages and to identify impacts of climate policies on global and regional economic systems.[3]

The top-down component consists of an inter-temporal optimal growth model, in which the energy input of the aggregate production function has been expanded to yield a bottom-up description of the energy sector. The model provides a fully intertemporal allocation of investments in energy technologies and R&D that is used to evaluate optimal and second-best economic and technological responses to different policy measures.

Countries are grouped in regions that cover the world and whose strategic interactions are modelled through a dynamic game. The game theory set-up accounts for interdependencies and spillovers across regions of the world, and equilibrium strategies reflect inefficiencies induced by global strategic interactions. This set-up allows the analysis of both fully cooperative equilibria (in the case in which it is assumed that all regions of the world sign a climate agreement) and partial/regional coalitional equilibria (when only a subgroup of regions signs the agreement or different groups of regions sign different agreements).

In WITCH, technological progress in the energy sector is endogenous, thus accounting for the effects of different stabilisation policies on induced technical change, via both innovation and diffusion processes. Feedbacks from economic variables to climatic variables, and vice versa, are also accounted for in the model's dynamic system.

3.2 BREADTH AND DEPTH OF INTERNATIONAL AGREEMENTS

The crucial issue for successfully addressing the problem of global warming is to enlist ambitious mitigation actions by a sufficient number of emitters. In this context, three aspects need to be considered: (i) the incentives for the main emitting countries to participate in the climate coalition. These depend on a wide range of factors, in particular avoided damage and abatement costs[4] for individual countries and for the group of countries undertaking mitigation; (ii) the identification of potentially effective coalition (PEC), i.e. coalitions of countries that cooperate on emission reduction with aggressive policies that enable the achievement of the desired GHG concentration target;

and (iii) the stability of such coalitions, i.e. the lack of incentives for individual countries to abandon the coalition (agreement) because their welfare would be larger when free-riding. Moreover, for coalitions which are not stable, the question arises as to whether there is a set of international (monetary) transfer designed to provide sufficient incentives to potential free-riders to join and remain in the coalition.

These issues are carefully examined in Bosetti et al. (2013) by applying the WITCH model to assess parties' incentives to participate, the environmental effectiveness of different coalitions, their inherent stability, and the available policy options to change the incentives of individual countries as well as improve the stability of coalitions.

(Dis)incentives to Cooperate

In the WITCH world, a region's decision to take part in a climate coalition is driven mainly by two factors: the damages it expects to suffer from climate change; and the costs of technologies to abate GHG. Both these factors are strongly influenced by the inter-temporal trade-offs, governed by regions' pure rate of time preferences.[5]

In WITCH, the damages suffered by regions are modelled through a reduced form climate damage function, linking temperature changes and gross world product. Despite a large and growing research effort started in the 1990s, significant uncertainties on future climate change impacts remain. In order to reflect this uncertainty, Bosetti et al. (2013) consider two alternative scenarios: a low-damage scenario, embedded in the basic version of WITCH, and a high damage scenario.

The results of the model are in line with evidence from the impacts literature: in both scenarios, damages are higher in developing countries – in particular Sub-Saharan Africa and South Asia (including India), driven mostly by higher damages in agriculture and the increase of vector-borne diseases (Sub-Saharan Africa) and catastrophic climate impacts (South Asia). Amongst developed countries, it is Western Europe that experiences the highest damages, again because of the impacts of climate change on agriculture, but also expected damages to coastal settlements, and climate extremes. On the other hand, small benefits would be expected in some regions of the world, such as China, Eastern European countries, Japan and Korea. The result is driven mostly by a reduction in energy demand for heating purposes or positive effects on agricultural productivity (in China). In the higher damage scenario, which defines an upper bound accounting for most of the wide and uncertain range of estimates proposed in the literature, damages from global climate change are about twice a large as in the low damage one. This is shown in Figure 3.1 and 3.2.

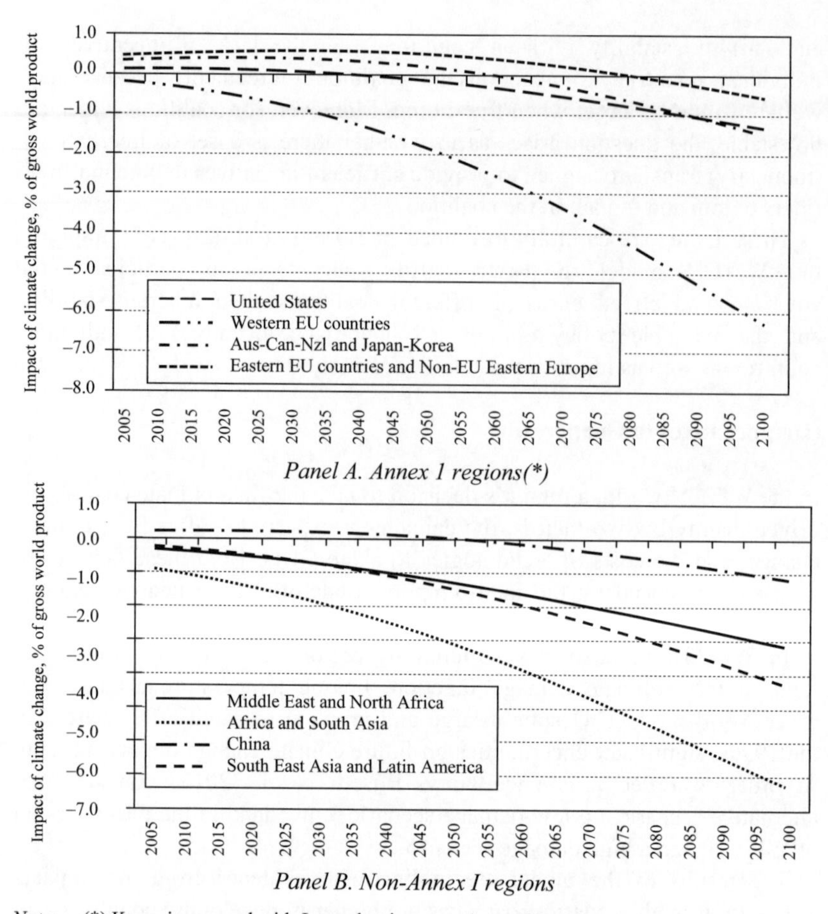

Note: (*) Korea is grouped with Japan, but is not an Annex I country.

Source: WITCH model simulations. See Bosetti et al. (2013).

Figure 3.1 Regional damage functions in the basic version of the WITCH model – baseline and high impact scenarios

The second element that is considered critical in determining whether mitigation actions will be undertaken is the cost of meeting emission reduction targets. Countries that face larger costs of abatement can expect to gain less from joining a climate coalition, and therefore have larger incentives to free-ride, other things being equal. Mitigation costs, measured in terms of discounted consumption loss from alternative world carbon tax scenarios, are found to be higher in developing countries, due to their higher energy/carbon intensity. Fossil fuels producers are the largest losers, due to

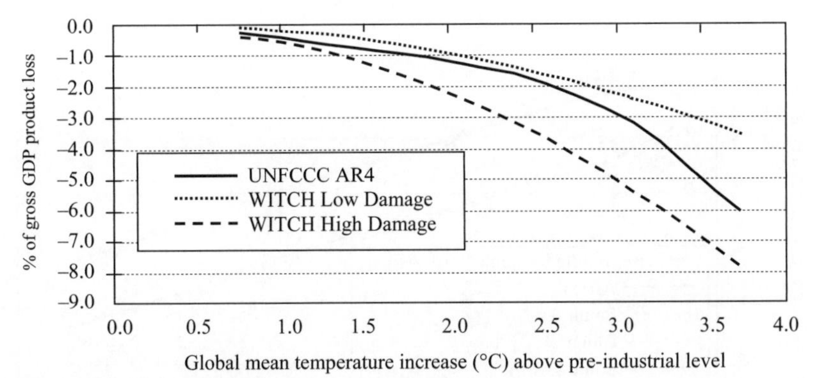

Source: WITCH model simulations. See Bosetti et al. (2013).

Figure 3.2 High damage scenario

both their high energy/carbon intensity and the fall in world fossil fuel prices (Figure 3.3) (see also Massetti and Tavoni, 2011).

From the modelling results, therefore, it emerges that developing countries face both higher damages from climate change and higher abatement costs. In light of these results, what are the incentives that could lead to a broad coalition of countries engaging in climate change abatement? Can global agreements that include developing countries, if any exist, be stable? If not, could informed policies change the balance of incentives? These questions are addressed in the following sub-sections.

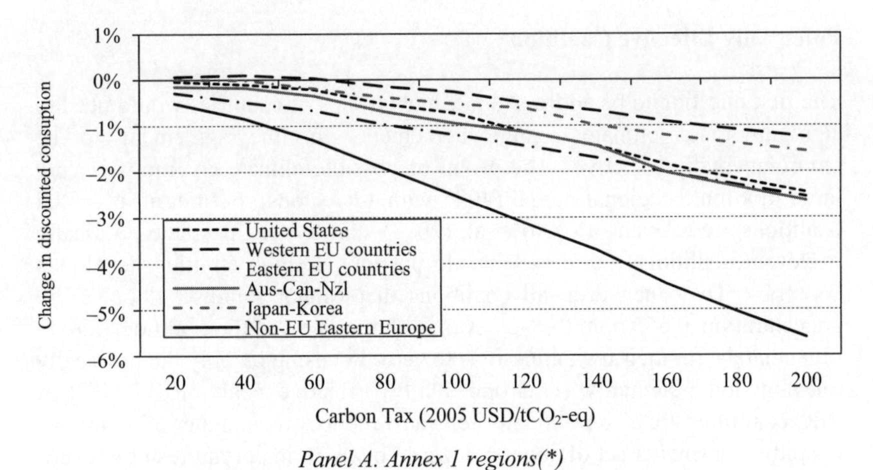

Panel A. Annex 1 regions(*)

Figure 3.3 Discounted regional abatement costs under a range of world carbon tax scenarios

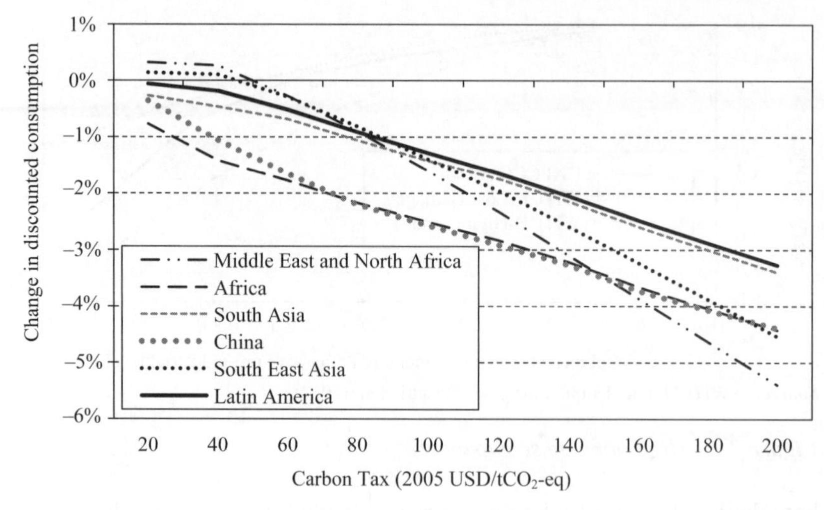

Panel B. non-Annex I regions

Notes: (*) Korea is grouped with Japan, but is not an Annex I country.
Cumulative consumption gap relative to baseline in present value terms over 2015-2100, using a three percent annual discount rate.

Source: WITCH model simulations. See Bosetti et al. (2013).

Figure 3.3 Discounted regional abatement costs under a range of world carbon tax scenarios (Continued)

Potentially Effective Coalition

The first question to be addressed is which groups of countries would need to sign up to a climate change agreement, for the agreement to be environmentally effective? The group of possible climate coalitions is very large in a multi-regional model (4095 with 12 regions), but not all potential coalitions are relevant. Bosetti et al. (2013) start by leaving aside politically irrelevant coalitions, i.e. those that do not include high emitting developed countries. They then drop all coalitions that cannot stabilise global GHG concentration at 550 ppm CO_2-eq even if all members of the coalition were to immediately cut their own emissions to zero. Focusing on only the politically relevant and potentially (environmentally) effective coalitions (PEC) (all PEC coalitions are also politically relevant) reduces the number of coalitions to study to a smaller set of about a dozen. This is an important result in itself: almost all world countries should agree to mitigation targets if the world emission path is to be consistent with a long-run 550 ppm CO_2-eq GHG concentration target.

The results are summarised in Table 3.1.The table considers two possible time-horizons for the climate agreement, 2050 and 2100, both consistent with the objective of keeping GHG concentration below 550 ppm CO_2-eq in 2100. If the time-horizon of the agreement is 2050, all PEC coalitions (11 in total) must include industrialised countries as well as China and India (Panel A). If the time-horizon is 2100, only three coalitions are PEC, and they include all world regions with only a few exceptions (Panel B). These findings hold under both high and low damage functions, as well as under high and low rates of time preferences.

Being potentially able to achieve the 550 ppm CO_2-eq GHG concentration target does not mean that the coalition will actually chose to achieve that concentration target. In order to check what is the endogenous, optimal, level of GHG concentrations, Bosetti et al. (2013) solve the WITCH model by grouping all regions that belong to the coalition into a single region – the coalition – and leave all regions that do not belong to the coalitions as independent entities – the free-riders, or singletons. The regions in the coalition choose the optimal GHG concentration level by jointly maximising welfare, i.e. by comparing the costs and benefits of reducing emissions (cost-benefit analysis).[6]

Results of cost-benefit runs that use a high damage function and a low intertemporal discount rate (the combination that yields the highest emission reductions) indicate that only the grand coalition – the coalition that includes all regions – chooses to achieve a level of GHG concentrations in 2100 that is close to 550 ppm CO_2-eq.[7] Table 3.2 below shows the results of the cost-benefit analysis for the six coalitions that exclude at least Sub-Saharan Africa, the poorest region of the world that will always be excluded from a partial agreement. The results are striking. Even if only Sub-Saharan Africa is excluded, it is not possible to keep global GHG concentrations below 600 ppm CO_2-eq in 2100, a level in itself not compatible with the 2°C warming limit.

Two forces are at play here: (i) the smaller the coalition, the less it internalises the environmental externality; and (ii) the smaller the coalition, the larger the number of regions that can set their emission freely and, therefore, free-ride on the coalition's effort. Consider, for instance, the case of the coalition including all regions but Sub-Saharan Africa: as shown in Figure 3.4, the optimal abatement path of the coalition will be lower, as the large expected damages suffered by Sub-Saharan Africa are not taken into account by the coalition members when setting their optimal abatement level. At the same time, Sub-Saharan Africa increases dramatically its emissions.

Table 3.1 Potentially effective coalitions to meet a 550ppm CO_2-eq target at the 2050 and 2100 horizons

Must participate	May not participate (Any combination of the following regions)
Panel A. PECs in 2050	
1. Developed countries, Latin America, Non-EU Eastern Europe (including Russia), South East Asia, Middle East and North Africa	Africa, South Asia (including India), China
2. Developed countries, Non-EU Eastern Europe (including Russia) China, Middle East and North Africa	Africa, South Asia (including India), South East Asia, Latin America
3. Developed countries, Non-EU Eastern Europe (including Russia) China, South East Asia	Africa, South Asia (including India), Latin America, Middle East and North Africa
4. Developed countries, China, South East Asia, Middle East and North Africa	Africa, South Asia (including India), Non-EU Eastern Europe (including Russia), Latin America
5. Developed countries, Latin America, China	Africa, South Asia (including India), Non-EU Eastern Europe (including Russia), Middle East and North Africa, South East Asia
6. Developed countries, Latin America, South Asia (including India), South East Asia, Middle East and North Africa	Africa, China, Non-EU Eastern Europe (including Russia)
7. Developed countries, Non-EU Eastern Europe (including Russia), South Asia (including India), South East Asia, Middle East and North Africa	Africa, China, Latin America
8. Developed countries, Latin America, Non-EU Eastern Europe (including Russia), South Asia (including India)	Africa, China, Middle East and North Africa, South East Asia
9. Developed countries, South Asia (including India), China, South East Asia	Africa, Non-EU Eastern Europe (including Russia), Middle East and North Africa, Latin America
10. Developed countries, South Asia (including India), China, Middle East and North Africa	Africa, Non-EU Eastern Europe (including Russia), South East Asia, Latin America
11. Developed countries, South Asia (including India), China	Africa, Latin America, South East Asia, Middle East and North Africa, Non-EU Eastern Europe (including Russia)

Table 3.1 (Continued)

Must participate	May not participate (Any combination of the following regions)
Panel B. PECs in 2100	
1. Developed countries, Non-EU Eastern Europe (including Russia), South Asia (including India), China, South East Asia, Middle East and North Africa	Africa, Latin America
2. Developed countries, Latin America, Non-EU Eastern Europe (including Russia), South Asia (including India), China, South East Asia	Africa, Middle East and North Africa
3. Developed countries, Latin America, South Asia (including India), China, Middle East and North Africa	Africa, Non-EU Eastern Europe (including Russia), South East Asia

Notes: A coalition is assumed to be a potentially effective coalition (PEC) at a given horizon if participating regions' lower bound emission levels, assumed here to be zero, added to the BaU emissions of non-participating countries (singletons) results in stabilisation of overall GHG concentration at 550ppm CO_2.eq or below.
Developed countries include Australia–Canada–New Zealand, Japan-Korea, United States, Western EU countries and Eastern EU countries.

Source: Bosetti et al. (2013).

Table 3.2 Analysis of the environmental achievements of potentially effective coalitions, cost-benefit mode, high-damage/low discount rate case

	Overall GHG concentration (ppm CO_2-eq)	
	2050	2100
Non-participating regions:		
None (Grand Coalition)	507	546
Africa	518	603
Africa, Latin America	532	612
Africa, Non-EU Eastern Europe	531	603
Africa, Middle East and North Africa	529	609
Africa, South East Asia	526	598
Africa, South East Asia, Non-EU Eastern Europe	529	603

Source: Bosetti et al., 2013.

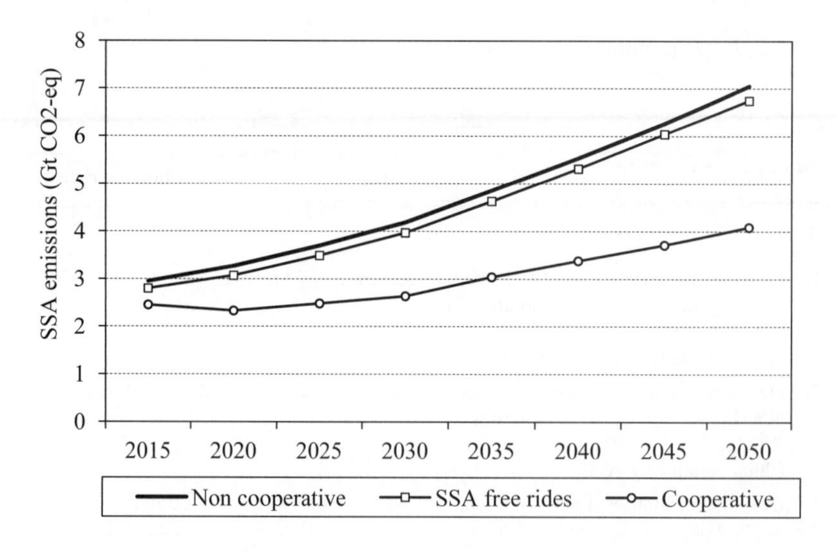

Source: Bosetti et al. (2013).

Figure 3.4 An illustration of free riding incentives: the case of Sub-Saharan Africa

Stable Climate Agreements

In addition to being environmentally effective, an international climate agreement needs to be struck in the first place, and it needs to be self-enforcing. In the game-theoretic setting, a region will decide to join a coalition (international climate agreement) only if it is profitable, and if it is stable. A coalition is said to be profitable if each signatory country has a higher welfare than in a scenario where the coalition is not formed (Business-as-Usual). A coalition is internally stable if signatory countries do not have incentives to defect and behave non-cooperatively when other members continue to cooperate. It is potentially internally stable if it can be turned to stability through a set of self-financed financial transfers across participating members. A coalition is externally stable if there is no incentive to enlarge the coalition by including non-signatory countries.

Table 3.3 reports profitability, internal and potential internal stability of a selected number of coalitions. As expected, all large coalitions considered in the analysis are not profitable for low discount rates, as there is always at least one region, which is worse off than in the non-cooperative case. In particular, China and Africa never benefit from a climate change agreement. It is interesting to note, however, that the grand coalition is profitable for a discount rate of 3 percent, as shown in the table. This in turn depends on the fact that when future benefits of avoided climate change are discounted at a higher rate, the coalition undertakes only moderate reductions. Hence, stability comes at the price of lower effectiveness.

Table 3.3 Profitability and stability of potentially effective coalitions

Non-participating regions	Profitability	Internal stability	Potential stability
Low PRTP (0.1%) – Negishi Weighted			
None (Grand coalition)	NO (Africa, China)	NOT STAB (All)	NOT PIS
Africa	NO (China)	NOT STAB (All)	NOT PIS
Africa, Latin America	NO (China)	NOT STAB (All)	PIS
Africa, Non-EU Eastern Europe	NO (China)	NOT STAB (All)	PIS
Africa, Middle East and North Africa	NO (China)	NOT STAB (All)	PIS
Africa, South East Asia	NO (China)	NOT STAB (All)	PIS
Africa, South East Asia, Non-EU Eastern Europe	NO (China)	NOT STAB (All)	PIS
High PRTP (3%) –Negishi Weighted			
None (grand coalition)	YES	NOT STAB (All but China and Latin America)	PIS

Note: Bosetti et al. (2013) use the so-called Negishi weights to aggregate welfare of different regions. These weights equate marginal utility of consumption across all regions, thus avoiding transfers from rich to poor regions. For further discussion and alternative weighting schemes see Bosetti et al. (2013).

The results shown in Table 3.3 also indicate that none of the coalition is internally stable, irrespective of the discount rate. Thus, at least one region sees incentives to defect and benefit from the improved climate. The free rider also benefits from lower fossil fuel costs because all participating countries reduce their consumption of coal, oil and natural gas, thus lowering their global and regional prices. On the other hand, while several coalitions are potentially internally stable – that is, the aggregate residual surplus of consumption in the coalition is greater than the sum of the discounted consumption gains that countries would have when they free-ride – the grand coalition is not. The aggregate, discounted surplus from cooperation is equal to USD477 trillions over the 2005–2100 time-horizon, while the sum of the gains individual regions would get when free-riding is equal to USD680 trillions. The coalition involving all regions except Africa – the second most environmentally effective coalition after the grand coalition – is also not potentially internally stable: the difference in discounted surplus from cooperation and non-cooperation, however, is much smaller than in the case of the grand coalition, equal to 2 percent of the aggregate discounted consumption gain of the coalition. Among the potentially internally stable coalitions, the most environmentally effective grouping would reach a GHG concentration slightly above 518 ppm CO_2-eq in 2050 and around 600 ppm CO_2-eq in 2100.

With a low discount rate, the incentives to free ride (given by the difference in intertemporal welfare per capita between free-riding on, and participating in, the grand coalition) are largest in the Middle East-North Africa region, China, the rest of Africa, and non-EU Eastern European countries. By contrast, developed countries have the lowest free-riding incentives, with the exception of the Australia–Canada–New Zealand region (Figure 3.5).

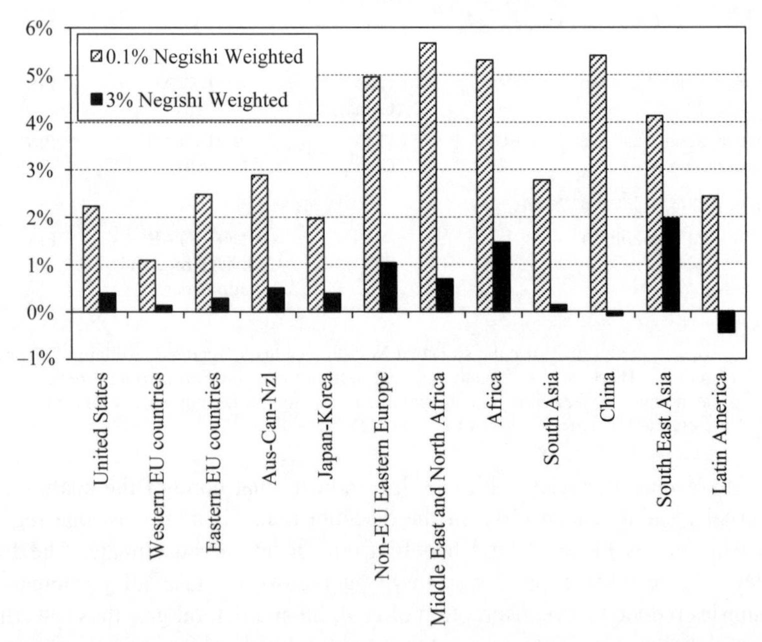

Figure 3.5 Estimated ranking of free-riding incentives across regions

As we have seen, with higher discount rate (3 percent) each region's welfare is larger in the grand coalition than in the non-cooperative case, i.e. the grand coalition is profitable, and the incentives to free ride are now much smaller. Despite this, the coalition is not internally stable. The discounted surplus, in consumption terms, generated by cooperation (USD133 trillions) is four times greater than the discounted aggregate surplus from free-riding (USD35 trillions).

The findings of this Section are in line with established results: when gains from cooperation are large, as in the 0.1 percent discount case, free riding incentives are also likely to be high. On the contrary, when gains from cooperation are small, as in the 3 percent case, free riding incentives are reduced significantly.

A clear message emerges from the results: cooperation is possible, profitable and potentially stable, but only if the environmental target is moderate (at least compared to what currently discussed in the policy debate), i.e. around 600 ppm CO_2-eq by 2100. Optimising behavior of members of the coalition could result in more ambitious targets with lower mitigation costs, higher climate change damages, or a combination of the two.

The results of this section are in line with the current developments of global negotiations on climate change, but they are subject to a number of limitations. First of all, the results are sensitive to modelling assumptions – in particular, there are significant uncertainties over future emission trends, the market and non-market impacts of climate change, or catastrophic events. Secondly, the co-benefits of mitigation actions, such as on human health, biodiversity, energy security, among others, are not taken into account. Recent analyses indicate that such co-benefits can indeed be significant (Bollen et al., 2009, Burniaux et al., 2008) and could, therefore, provide added incentives for cooperation. Thirdly, the removal of environmentally harmful subsidies, particularly fossil fuels, is not considered. Such a policy could improve incentives to participate in mitigation action, by yielding economic gains and lowering the carbon intensity of economies. Fourthly, the absence of negative emissions technologies or any other technology that might alter the climate (i.e. geo-engineering), and the absence of adaptation policies, are also two limitations of this analysis. From a policy perspective, the welfare maximisation paradigm may not be applicable to international negotiations. Current international redistributive policies, such as official development assistance, seem to indicate the existence of some degree of altruism. Therefore, some countries may, in reality, sign up to a binding agreement even if, in principle, they would stand to gain more from free-riding. The analysis focuses on immediate, irreversible, and self-enforcing participation in mitigation action, thereby abstracting from other possible bargaining strategies, such as delayed participation, renegotiation, or sanctions. Finally, the analysis in this section has focused on assessing incentives for countries to undertake joint mitigation action in the absence of an external international agreement. The resulting emission path is well below what would be required to meet the 2°C target, called for by the scientific community to avoid irreversible damages and tipping points.

The next section explores more in details some of these elements, in particular focusing on the impacts of delayed actions and uncertainty over the feasibility and costs of achieving an emission path consistent with the target of keeping temperature increase below 2°C.

3.3 DELAYED ACTION

Even if actions to counteract climate change seem to be warranted at some level, there is still little agreement over what the ultimate target should be, as the long-term stabilisation target is clearly a political decision. Despite optimisms, uncertainty for policy makers remains a critical issue. It is thus of paramount importance to understand whether expectations of gaining better information should induce policymakers to wait before acting, or, conversely, to act promptly before it is too late. Bosetti et al. (2008b and 2009b) adopt a stochastic optimisation approach to explore the impact of uncertainty on the strategic behavior of players. It is important to point out that the analysis shifts away from the joint identification of the optimal emission level and its related mitigation costs, focusing rather on the cost of achieving an exogenously determined emission path. This recognises the fundamentally political nature of international climate agreements.

A Closing Window of Opportunity?

It is often argued that the many uncertainties surrounding climate change, its impacts and mitigation costs, warrant a delay in taking expensive mitigation actions. Indeed, the results of earlier papers seem to indicate that the costs of reducing emissions are lower if the majority of the efforts are deferred into the future (see, for instance, Wigley et al., 1996, Richels and Edmonds, 1995, Kosobud et al., 1994). Since these initial works, much effort has been devoted to identifying the optimal timing of abatement, focusing in particular on the impacts of uncertainty and learning on policy decisions: even though the results are not unequivocal, they seem to indicate that, in general, the prospect of learning new information would lead to a reduction in current abatement (see Ingham et al., 2007, for a review of the literature).

However, much less effort has been devoted to assessing the implications of having to deviate from the chosen emission reduction path at some point in the future. Delaying action may significantly restrict the options available to lower GHG concentration in the future – thus increasing the risks of severe climatic shifts. The slow process of GHG accumulation and decay may imply that, if we fail to take action to reduce emissions now and we find out at a later date that climate change is more serious than expected, we may no longer be able to reverse the climatic changes that have been triggered, no matter how stringent a policy we implement. By delaying action in search for more certainty we may miss the current window of opportunity to take decisive action against climate change.

Bosetti et al. (2009b) explore the economic costs associated with a waiting strategy – in particular, what would the cost be of continuing along the BaU path for 20 years, and agreeing on some stabilisation target at a later date?

The analysis is computed by assuming that a global cap-and-trade scheme[8] is implemented either now or in 20 years time, in order to achieve an agreed stabilisation target (either 450 or 550 ppm CO_2 only). The optimal investment paths – in particular in the energy sector – are computed and the economic cost of the climate policy is estimated.[9] Table 3.4 summarises the main results.

If policy makers decided to take immediate action now, the net present value (NPV) of stabilisation costs, evaluated at a 3 percent (5 percent) discount rate, would range from 0.3 percent (0.2 percent) to 3.5 percent (2.3 percent) of Gross World Product (GWP), depending on the stringency of the target. These represent the cost of stabilisation as GWP losses compared to the BaU scenario. As the WITCH model explicitly accounts for free-riding incentives and other inefficiencies deriving from environmental, R&D, and market externalities, these estimates are somewhat higher than the figures proposed by the IPCC Fourth Assessment Report (see Bosetti et al., 2007). The cost of getting the short term policy wrong and then adjusting it, as shown in Table 3.4, varies depending on how stringent later action has to be. In particular, moving from the BaU to a stringent climate stabilisation target after the 20-year delay is extremely costly-up to 7.6 percent (5.5 percent at 5 percent discounting) of the net present value of GWP over this century. This is equivalent to an increase of policy costs of about 130–140 percent, and amounts to an annual loss of 5.7 (2.2 at 5 percent discounting) trillion USD per year of delay. The magnitude of these losses is well beyond what is commonly held by policy makers to be an economically feasible stringent climate change policy.

On the other hand, the costs of delaying to undertake mild actions are relatively modest (the net present value of the cost is equivalent to 0.4 percent of GWP as opposed to 0.2 percent of GWP with an immediate start).

This leads to another important result of the analysis: in line with the precautionary principle, a policy strategy that immediately begins to undertake some emissions reductions that are consistent with a 550 ppm CO_2 only stabilisation target and, in 20 years, reverts to the BaU scenario, does not harm global welfare. On the contrary, the analysis suggests that this

Table 3.4 The cost implications of delayed action, NPV GWP loss to 2100, discounted at 3 percent (5 percent)

	Continue along the Business-as-Usual path	Undertake a mild climate control policy(a)	Undertake a stringent climate control policy(b)
Take action now	.	0.3% (0.2%)	3.5% (2.3%)
Wait 20 years on the Business-as-Usual path	.	0.4% (0.3%)	7.6% (5.5%)
Wait 20 years on a mild policy(a) path	–0.03% (0.06%)	.	4.2% (2.7%)

Notes: (a) 550ppmv CO2 only, roughly equivalent to 650ppmv all gases.
(b) 450ppmv CO2 only, roughly equivalent to 550ppmv all gases.

Source: Bosetti et al. (2009b).

strategy would actually lead to a very mild increase of GWP, thanks to the internalisation of market failures and externalities, in particular those related to carbon, exhaustible resources and innovation. If a tighter mitigation is needed in 20 years, shifting from a mild to a stringent target is less costly than continuing along a BaU path for the next 20 years, with the net present value of the costs reaching 4.2 percent (2.7 percent) of GWP, still well below the cost of inaction for 20 years followed by embracing a stringent climate policy in 20 year's time.

The policy implications of this exercise are quite clear, and support the arguments that call for immediate action to tackle climate change: if we continue doing nothing for 20 years, the costs of shifting from a BaU to a stringent climate policy are extremely high. On the other hand, undertaking some form of mild stabilisation policy seems to be a hedging strategy which, at virtually no cost, would allow us to revert to BaU and, at relatively modest cost, to undertake more decisive action if a more stringent stabilisation policy is decided upon. Committing to a mild mitigation effort in the short-term significantly decreases the costs of delaying stringent action – while, at the same time, increasing the probability of reaching a stringent agreement. A precautionary behaviour would thus both reduce the costs of stabilisation, and improve the chances of an international agreement consistent with what scientific evidence calls for.

Optimal Mitigation Policies under Climate Stabilisation Target Uncertainty

Another interesting question for policy makers is the identification of the optimal mitigation policies, when there is uncertainty over the stabilisation target that may be agreed upon in a future international climate regime. We have seen that the Cancun Agreements fall short of specifying a greenhouse concentration target – yet we can expect that, in the future, such a target will be agreed upon at the international level. What would then be the optimal strategy that policy makers should follow today, given the uncertainty over the mitigation target? For this exercise, a stochastic programming version of the WITCH model is used, where the uncertainty is characterised as a scenario tree. The scenarios are solved simultaneously, accounting for non-anticipativity constraints (action has to be the same for different scenarios before the disclosure of uncertainty, while the optimal reaction to the information revealed when uncertainty is eliminated is allowed afterwards). This modelling strategy provides a risk management approach that can help policy makers in the choice of a hedging strategy in the absence of a long-term target and in the face of significant uncertainty over the cost of mitigation, adaptation, and the impacts of climate change. This risk management approach is particularly relevant for investment strategies in the energy sector, where inertia in the system means that decisions taken today will have strong implications for future actions.

In these simulations, it is assumed that uncertainty is resolved in 2035, when three alternative scenarios might succeed, each with equal probability:[10]

- BaU (no target), where the world fails to agree on an emission reduction target, and countries continue along the Business-as-Usual path.
- 550ppm CO_2 stabilisation, where an international agreement is reached with a target of carbon concentration in 2100 of 550 ppm CO_2 only (650 all GHGs).
- 450ppm CO_2 stabilisation, where an international agreement is reached with a target of carbon concentration in 2100 of 450 ppm CO_2 only (550 all GHGs).

The optimal investment path for all energy technologies, physical capital and for R&D is computed as a non-cooperative game, where each region behaves strategically.

If policy makers, in taking mitigation decisions today, take into account the uncertain results of international negotiations over a mitigation target, the optimal path of fossil fuel emissions is below the optimal path that would be chosen by a myopic decision-maker who does not take this uncertainty into

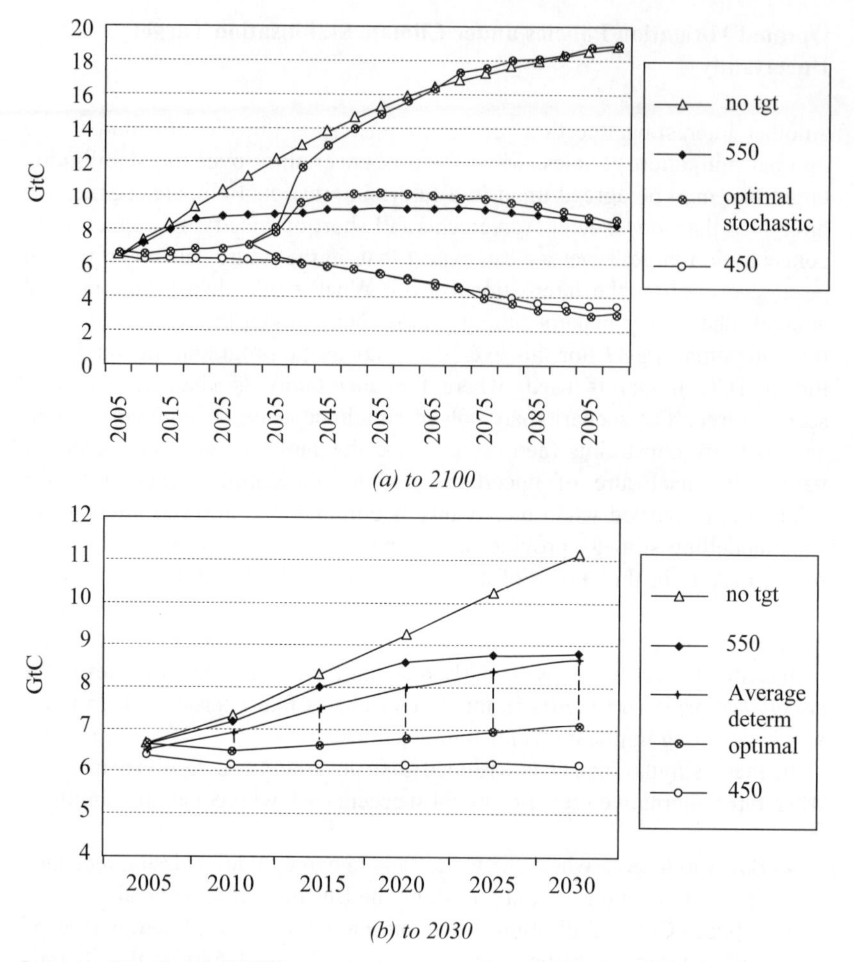

Source: Bosetti et al. (2009b).

Figure 3.6 Fossil fuel emissions for optimal stochastic and deterministic case

consideration. The optimal strategy before the uncertainty is resolved, thus to engage in significant mitigation action, providing a clear indication of hedging behaviour. In 2030, emissions are 57 percent lower than in the BaU scenario: when there is equal chance of facing no climate policy, a mild climate policy or a stringent climate policy in 30 years' time, the best short-term strategy is to minimise the growth in emissions. Interestingly, if we compare the optimal emission path under uncertainty with the optimal path under a mild and stringent mitigation policy starting today, we find that the optimal strategy is to reduce emissions 23 percent below the path under a mild policy, and only 14 percent higher than in the case of a stringent policy. This result is driven by the possibility of a stringent climate target in the future – albeit this only has a 33 percent change of occurring. The optimal emission path is always below the average of the three deterministic cases (see Figure 3.6b) by 22 percent: This quantifies the dimension of the precautionary strategy.

It is reasonable to expect that there are economic costs in ignoring the uncertainty surrounding the mitigation target in the future when taking investment decisions today. An understanding of these costs and their magnitude could facilitate the achievement of an agreement in the future.

In order to do this, Bosetti et al. (2009b) run the stochastic version of the model, but fix all the choice variables to those of the three deterministic cases until 2035 (the time frame to uncertainty resolution); from that point on, the model is free to optimise for each of the three states of the world, given the sub-optimal choices undertaken during the preceding periods. This enables the quantification of the economic losses that would result from sticking to a BaU scenario, a 550 ppm stabilisation policy or a 450 ppm stabilisation scenario (thus not adhering to the optimal strategy) in the periods before the final outcome of the international agreement on climate is known.

The first important result is that following the BaU scenario from now to 2035 does not allow a feasible solution to the optimisation problem if the target after 2035 is the ambitious one (450 CO_2 (550e)). In other words, a BaU strategy for the next three decades precludes the world from reaching an agreement on climate change that entails a stabilisation target of 550 CO_2 only.[11] This result extends the one seen in the first part of this section, which showed that policy costs increase sharply with the period of inaction.

As shown in Figure 3.7, a higher mitigation level results in economic losses initially, but leads to higher output after the resolution of uncertainty. This result is explained by the fact that, thanks to mitigation actions undertaken before an international agreement is reached on climate change, the burden of meeting the stringent target in the future has been alleviated. On the other hand, a sub-optimal 450 strategy inflicts costs that are somewhat higher than the benefits of a milder choice (−0.8 percent GWP losses in 2035 for 450, as opposed to +0.45 percent gains for 550) in the short-term. The picture reverses after 2035, when the costs of under-abatement before shifting to a stringent policy are higher than the benefit of a more virtuous early strategy.

It is clear that choices over discounting GWP over time determine the merit order of either sub-optimal strategy, as Table 3.5 shows. With a low discount rate (3 percent), the sub-optimal 550 ppm strategy entails an economic penalty of around 17 percent (from the 1.4 percent GWP loss for

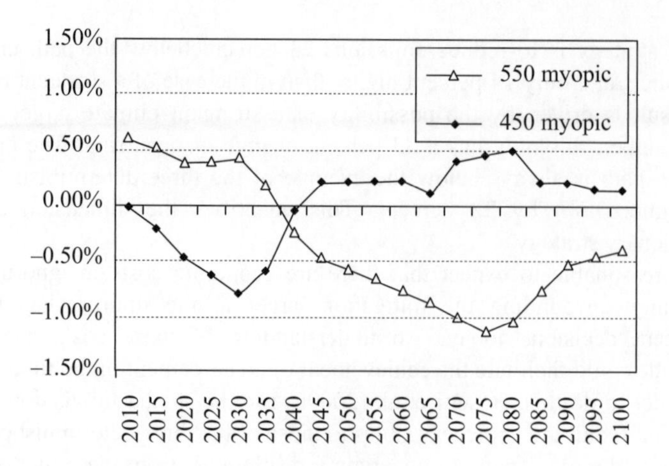

Source: Bosetti et al. (2009b).

Figure 3.7 GWP differences of sub-optimal strategies with respect to the optimal hedging strategy

Table 3.5 Economic penalty (extra NPV GWP loss) of sub-optimal strategies at 450 and 550 (with respect to the optimal strategy under uncertainty)

Discount rate for NPV calculations	3%	5%
Sub-optimal 550	17.20%	0.01%
Sub-optimal 450	1.70%	16.20%

Source: Bosetti et al. (2009b).

the stochastic case to 1.7 percent GWP loss in the myopic one), whereas in the 450 ppm case, the loss is much smaller (1.7 percent, from 1.4 percent to 1.5 percent). The opposite holds with a higher discount rate (5 percent): a milder climate change strategy has almost no negative economic impact on policy costs, as it allows for greater initial economic growth, whereas a more ambitious one leads to a penalty of about 16 percent.

Summing up, the optimal strategy before 2035 in the presence of uncertainty about future stabilisation targets is a mitigation policy that results in an emissions trajectory that lies between the 450 and 550 trajectories. Choosing either of those two policies results in economic inefficiency ranging from about zero to 16–17 percent of GWP, depending on the discount rate. Pursuing less mitigation than the 550 ppm policy significantly increases these extra costs, leading to the impossibility of meeting the climate objective in the case of zero early mitigation action. In this case, therefore, it would be impossible for world leaders to reach a climate change agreement that would entail stringent stabilisation targets, consistent with the temperature limit of 2°C contained in the Cancun Agreements. It would also thus be impossible to achieve an emission path consistent with a temperature limit of 1.5°C, called for by Small Island Developing States.

3.4 INTERNATIONAL COOPERATION WITH PHASED-IN PARTICIPATION

While the cost implications of delaying global actions are much higher than one would expect, in particular when the uncertainty surrounding any future global stabilisation target is taken into account, the participation rate of developing countries is another critical element in a global agreement over controlling climate change. This will determine the economic acceptability and political feasibility of achieving an agreement over a feasible stabilisation target that is compatible with scientific evidence.

Indeed, one of the problems that negotiators are facing in achieving an agreement over climate change mitigation targets and their burden sharing is the substantial disparities that exist in lifestyles in different parts of the world. The right to develop should not be compromised, as the argument goes, and developing countries are thus unwilling to undertake mitigation efforts unless paid for by Annex I countries (A1).

Yet, when we take a closer look at the emission levels of some major emerging economies in Non-Annex I countries (NA1), such as China or India, we soon realise that in the last decade absolute emissions from emerging economies have grown significantly, and today China is the largest emitters in the world, with emissions 10 percent higher, as shown in Table 3.6.[12] While the picture of emissions per capita is necessarily different, it is clear that abatement actions by developing countries – in particular emerging economies – are a necessary condition for an environmentally effective climate change agreement.

Despite fast growing emissions from emerging economies, the Bali Action Plan reaffirms the principle of common but differentiated responsibility and respective capabilities, thus emphasising the different role that Annex I and Non Annex I countries would play in an international climate agreement. When and how NA1 countries will participate in an international agreement is, however, not clear.

Table 3.6 China vs United States – emission levels

	China/US ratio	
	Total emissions	Per capita emissions
1992	0.48	0.10
1997	0.55	0.12
2007	1.13	0.26
2030	1.75	0.44

Source: Bosetti et al. (2009c).

Most estimates of the economic cost of climate policies assume the complete and immediate participation of major economies. Yet, reality may be quite different, as NA1 countries are likely to join an international treaty at a later date.

Using the WITCH model, Bosetti et al. (2008c and 2009c) assess the implication of NA1 countries joining an international treaty only in 2035. They show that, if developing countries' participation is delayed, the cost of stabilising at 450ppm CO_2 only would be much higher in terms of GWP losses – an increase of a 160 percent by 2030 and of 77 percent by 2050 with respect to the case in which all countries take immediate action (Figure 3.8). This is because the cheapest abatement options are found in developing regions, and under a delayed participation regime cannot be exploited. Furthermore, developing countries' delayed participation entails that long lifetime investments in infrastructures are made without taking into account future climate change policies, which result in prolonged fossil fuel dependency and, therefore, economic penalties.

The results of this analysis point to much higher cost of climate policy if developing countries do not participate immediately in the global effort to curb GHG emissions. The policy implication is that as many regions as possible – and in particular fast growing countries like China, India, Brazil and Russia – should be given incentives to join an international climate policy agreement right at the outset. Carraro and Massetti (2012) illustrate a well-defined pathway to include China and India in a global climate agreement.

One of the incentives that could prove useful in this respect is allowing these countries to participate in a global carbon market – trading emission reductions below their forecast BaU emission path, rather than below a binding emission reduction target.

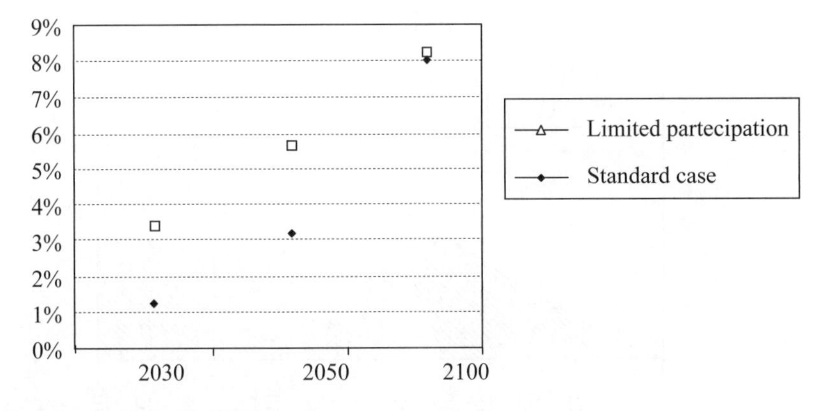

Source: Bosetti et al. (2009c).

Figure 3.8 GWP losses for a 450 ppm CO_2 only stabilisation target with delayed developing countries' participation

This policy option is explored in Bosetti et al. (2008c). When developing countries are allowed to trade emission reductions below their projected BaU emission levels, the cost of delaying their participation in the binding agreement is significantly reduced. Assuming, as in the previous case, that NA1 join the coalition in 2035, Bosetti et al. (2008c) explore the cost implications of different carbon trading regimes. Trading allowances are shared among world regions based on a 'contraction and convergence'[13] scheme, implying that emission allowances for the global trading scheme are initially allocated based on current emissions, though they gradually converge to equal per capita emissions by 2050. Under this scheme, NA1 would face a roughly constant target, if they were to join the agreement immediately. On the other hand, A1 countries would face a gradually decreasing target. Banking is permitted, to allow participatory countries to vary their emissions in response to other countries' delay in participation with respect to the global cap. However, in line with the Kyoto protocol rules, borrowing and speculative behaviour are not allowed.

With delayed participation of NA1, the mitigation target for A1 countries tightens very rapidly, to almost zero emissions in only 30 years (see Figure 3.9b).

Similarly, under a cap-and-trade regime where only signatory countries can participate, the price of carbon significantly increases, by roughly three times as compared to the scenario in which all countries take immediate emission abatement actions. In 2030, before NA1 join the coalition, the price of carbon reaches 1500 USD/tCO_2.

On the other hand, if NA1 countries were allowed to trade emission reduction below their projected BaU before they agree to binding emission

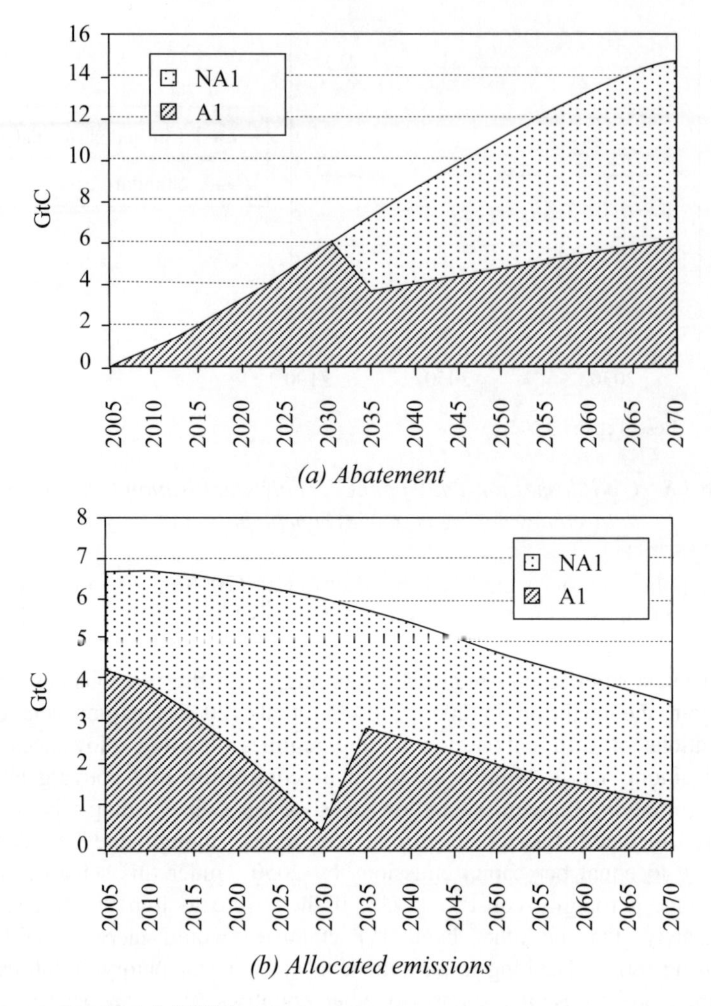

Source: Bosetti et al. (2008c).

Figure 3.9 Abatement and allocated emissions in the later participation scenario (2005–2070)

reduction targets in 2035, carbon prices would not significantly increase with respect to the immediate participation case. This is shown in Figure 3.10.

The strong effect of limiting participation in the carbon market to signatory countries is reflected in the global costs of achieving the 450 ppm CO_2 only mitigation target. Delaying participation of NA1 in the mitigation effort would entail a loss of GWP of 2.8 percent and 3.7 percent, depending on whether NA1 anticipate their future target or not. This implies an

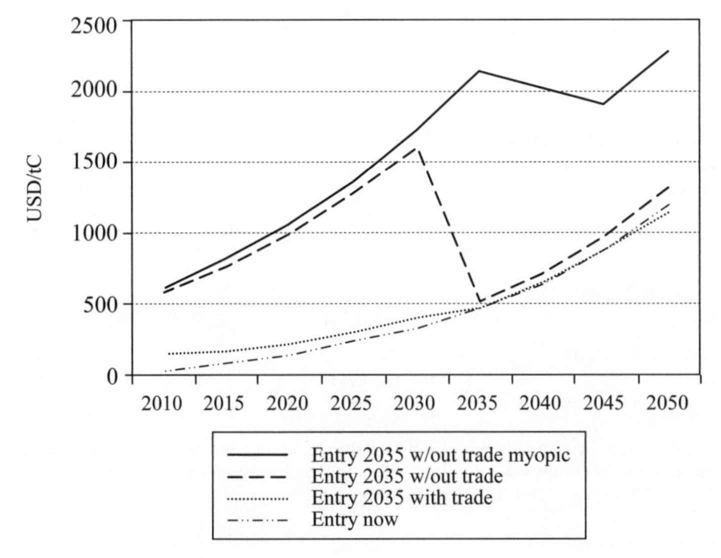

Source: Bosetti et al. (2008c).

Figure 3.10 Carbon prices across scenario

inefficiency in terms of reduced 'where-when' flexibility in the range of 11–25 trillion 1995 USD. However, if NA1 countries are allowed to trade immediately, the global costs are reduced, and are even below the scenario with immediate and full participation of NA1 – this result is driven by the larger incentives that A1 countries have in investing in R&D, which creates positive intertemporal spillovers on low-carbon technology. The results are summarised in Table 3.7.

As shown in Table 3.7, allowing NA1 countries to participate in a global carbon market even if they do not face binding reduction targets also makes sense from the perspective of A1: when NA1 are allowed to trade, the net present value of the abatement cost of A1 countries is reduced from 3 percent of their gross product to two percent – a reduction of about 10 trillion USD. Even for NA1, the scenario is more positive in the case where they are allowed to trade – with a cost in terms of losses of gross product falling from 2.5 percent to 2 percent. Global policy costs are shared proportionally by the two groups (both pay 2 percent of their GDP). Even if this would imply an additional transfer from A1 to NA1 of 3 trillion, it would save A1 a 13.9 trillion loss to achieve the same climate objective.

In conclusion, it is highly unlikely, from a political perspective, that NA1 would accept binding emission reduction targets immediately. At the same time, they will need to participate in the global mitigation effort if

Table 3.7 Policy costs across scenario – delayed participation, with and without trade

NPV GDP Loss 2000–2100	World	A1	NA1
Entry now	2.10%	1.70%	2.80%
Entry 2035 w/o trade	2.80% (+11 USDTln)	3.00% (+12.6 USDTln)	2.50% (–1.6 USDTln)
Entry 2035 w/o trade and myopic*	3.70% (+24.6 USDTln)	3.10% (+13.9 USDTln)	5.10% (+10.7USDTln)
Entry 2035 with trade	2% (–0.9 USDTln)	2% (+3 USDTln)	2% (–3.9 USDTln)

Note: In the myopic case, NA1 countries cannot anticipate that they will agree on a stabilisation target from 2035 onward. Therefore, all their choice variables are fixed at the BaU levels from 2005 to 2035. This case is called 'w/out trade myopic'.

Source: Bosetti et al. (2008c).

environmental integrity is to be ensured. Using WITCH, the implication of delaying binding targets for developing countries is explored: the analysis shows that the difficulty and costs of meeting a 450 ppm CO_2 only stabilisation target significantly increase when NA1 do not participate in a binding agreement. However, allowing NA1 to participate in the global carbon market by selling emission reductions below their projected BaU, would significantly reduce the costs of climate policy, while at the same time increasing financial transfer to NA1. It may therefore be a policy compromise that would make global concerted action against climate change more feasible – a suggestion in line with existing literature on international agreements (Weyant and Hill, 1999).

3.5 ALTERNATIVE REGIMES – BROADENING PARTICIPATION

The analysis in the previous sections of this chapter has shown that, without the concerted effort of all regions, either the globally optimal mitigation target would be low, or the cost of achieving a given concentration level would escalate. Therefore, addressing climate change requires both deep and broad international agreements. Most of the analysis surveyed in this Chapter has assumed that marginal abatement costs are equated in both rich and poor economies. This guarantees efficiency, and ensure a uniform carbon price at the global level. However, it does not necessarily imply equity. Historical responsibility for emissions, as well as the Rio principles of common but differentiated responsibility and respective capabilities, refer specifically to equity issues and call for differentiated targets and levels of effort. Introducing equity issues in climate negotiations further complicates the puzzle, but adds a more realistic dimension to the analyses. There is therefore a growing body of literature examining a large set of 'international climate architectures' that creatively expand the benchmark case of global, instantaneous participation with uniform carbon pricing to better take into account equity issues (Aldy and Stavins, 2007).

This section reviews work done with WITCH to quantitatively compare the main architectures for agreement proposed in the literature, based on four criteria:[14] economic efficiency; environmental effectiveness; distributional implications; and political acceptability, measured in terms of feasibility and enforceability. The ultimate aim is to derive useful policy implications that could provide insights for designing the next agreement on climate change.

Architectures for Agreement

Eight policy architectures, which have been discussed in the literature or have been proposed as potential successors to the Kyoto agreement, are assessed against a scenario without climate policy (BaU). These architectures are inspired by the proposals put forward within the Harvard Project on International Climate Agreements,[15] and are summarised in Table 3.8. The focus is on CO_2 mitigation only, excluding other greenhouse gases. Furthermore, as climate leaks and free riding incentives are likely to be substantial for less than a global agreement, all the proposed architectures envision that countries commit to at least not exceeding their projected emissions under the BaU scenario. This modelling assumption allows countries to trade in a global carbon market, undertaking cheap abatement and receiving financial resources for it, thus significantly lowering the costs of climate policy – and increasing its economic feasibility.

Following from the previous analysis, it is clear that scope and timing of mitigation policies are the key factors that determine the political and economic feasibility of an international climate policy regime. Universal agreements involve all regions, while partial agreements only require cooperation among a subset of regions. Agreements may require immediate efforts from participating countries, or they may take into account differential abilities to undertake abatement and, therefore, involve incremental participation, where some regions – usually transition economies and developing countries – are allowed to enter the agreement at a later point in time, when they satisfy some pre-defined criteria. A further distinction across architectures is the type of policy instrument involved: most schemes use a

Table 3.8 Architectures for agreement

Name	Key feature	Policy instrument	Scope	Timing
CAT with redistribution	Benchmark cap and trade	Cap and trade	Universal	Immediate
Global Carbon Tax	Global tax recycled domestically	Carbon tax	Universal	Immediate
REDD	Inclusion of REDD	Cap and trade	Universal	Immediate
Climate Clubs	Clubs of countries	Cap and trade and R&D	Partial	Incremental
Burden Sharing	Delayed participation of developing countries	Cap and trade	Universal	Incremental
Graduation	Bottom up targets	Cap and trade	Partial	Incremental
Dynamic Targets	Political feasibility	Cap and trade	Universal	Incremental
R&D Coalition	R&D cooperation	R&D	Universal	Immediate

Source: Bosetti et al. (2009d).

cap-and-trade approach, but carbon taxes and R&D policies are also considered. The eight architectures can be summarised as follows:

- *Global cap-and-trade with redistribution*. This is a benchmark scenario representing the first-best world, where all countries participate immediately in a global cap-and-trade system designed to stabilise atmospheric concentration of CO_2 at 450ppm by 2100. Tradable permits are allocated to all countries on an equal per capita basis.
- *Global tax recycled domestically*.[16] All countries apply a globally consistent carbon tax designed to achieve the same stabilisation trajectory as the previous scenario. Revenues from the tax are recycled domestically, and implementation begins immediately.
- *REDD (Reducing emissions from deforestation and forest degradation)*. Same as the first scenario, except credits from avoided Amazon deforestation[17] are included in the permit market.
- *Climate clubs*.[18] A group of mostly advanced economies agrees to abide by its Kyoto target and reduce CO_2 emissions 70 percent below 1990 levels by 2050. Other fast growing countries and regions (China, India, Latin America, Transition Economies, and the Middle East) begin gradual efforts to reduce emissions below their BaU,[19] but

converge to the same level of reductions as the first group of countries after 2050. All remaining countries face no binding targets, but their emissions are limited to the BaU.

- *Burden sharing.*[20] Annex I countries commence abatement immediately, with the burden sharing on an equal per capita basis. Binding emission reduction targets are extended to all countries, except Sub-Saharan Africa, in 2040.
- *Graduation.*[21] Countries adopt binding emission reduction targets as they reach specified bottom-up criteria for income (capability) and emission (responsibility). The first graduation step is reached when the average of the two criteria is satisfied, that is, emissions per capita match the average world emissions per capita, and income per capita increases to USD5000 (2005 value). When countries reach the first graduation level, their abatement target is equivalent to 5 percent with respect to 2005 emissions. The second graduation period is reached when emissions per capita are 1.5 of the world's average and income per capita is USD10,000. It entails a reduction in emissions by 10 percent with respect to 2005 levels. The only exception is China, which reduces emissions gradually, starting from 2050, in order to cut the level by 50 percent with respect to the 2005 emissions. Sub-Saharan Africa never graduates, and therefore faces no binding targets. It commits, however, to the BaU, so that it can participate in the carbon market. Annex 1 countries compensate for the delayed entry of NA1 by undertaking additional reductions required to achieve stabilisation at 450ppm CO_2 only by 2100.
- *Dynamic targets.*[22] Different countries adopt different targets over time depending on current and projected emissions (responsibility), income (capability), and population. The bottom-up targets are based on progressive cut factors – with respect to emissions in 1990 for the first period, and then with respect to projected emissions in the BaU scenario, corrected by a Lieberman-Lee Latecomer Catch-up factor for countries that have not yet ratified the Kyoto Protocol. Progressive cut factors take into account historic emissions relative to the emissions of the EU in 1990, current and projected emissions in the BaU, income per capita relative to the EU average, and population. Targets are defined for all regions, with the world divided into three broad groups: early movers (Europe, US, Canada, Japan, New Zealand, Australia, South Korea, and South Africa) take action from 2010–2015–2025; the late comers, with China and Latin America facing binding emission reduction targets from 2035, and India from 2050; and all other regions, that agree not to exceed their emissions under the BaU scenario and can thus take part in the international market for carbon permits. Sub-Saharan Africa does not face any emission target until

2030 – after which it enters the market for carbon permits by committing to not exceeding its BaU emissions.

- *R&D and technology development.*[23] This architecture is very different from the others, in that it entails no binding emission targets. Instead, all countries contribute a fixed percentage of their GDP to an international fund for developing low-carbon technologies. The share is roughly equal to the double of public energy R&D expenditure in the 1980s, which is 0.2 percent of regional GDP. Financial resources are redistributed to all regions on an equal per capita basis, and they are equally split to foster deployment of two key low-carbon technologies – wind and solar and carbon capture and storage – and for innovation in a breakthrough zero carbon technology in the non-electric sector.

Climate Effectiveness

The first and most important objective of a climate treaty is its environmental integrity. The implications of the eight architectures for agreement described above, in terms of industrial CO_2 emissions, are represented in Figure 3.11.

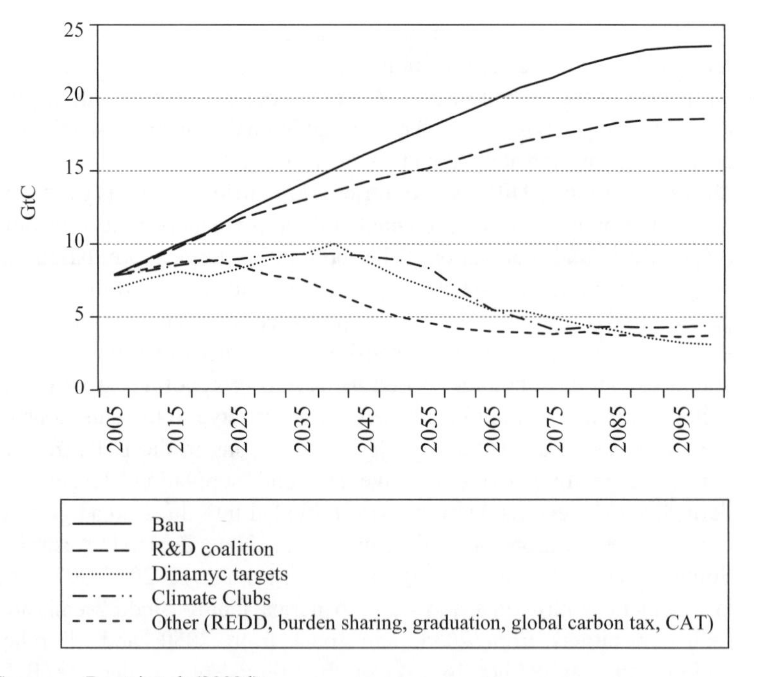

Source: Bosetti et al. (2009d).

Figure 3.11 Global energy CO_2 emission paths

As expected, the broader forms of agreement (the three global coalitions with an explicit target of atmospheric concentration of CO_2, implemented with different instruments) lead to a stabilisation of emissions at well below 5GtC by the middle of the century. Interestingly, one of the architectures based on bottom-up targets – the Graduation architecture – also leads to the stabilisation of CO_2 concentrations at 450 ppm.

The same level of emissions is achieved by the other agreements with incremental participation, but through a different transition path, reflecting the different dates at which regions start to face binding constraints. The implied concentration of atmospheric CO_2 is less stringent, stabilising at about 550 ppm. Finally, the global coalition cooperating on energy R&D does not achieve the stabilisation of CO_2 emissions, though the lower cost of low-carbon technology does lead to lower emissions than the BaU.

Ultimately, the environmental integrity of an international climate change policy should be assessed on the basis of the implied temperature changes with respect to pre-industrial levels.[24] These are reported in Figure 3.12. Though the exact magnitude of temperature changes is uncertain, the figure is helpful in comparing across the various architectures. In the BaU scenario, where no international policy to curb CO_2 emissions is implemented, temperature change is expected to reach 3.7°C above pre-industrial levels in 2100. When cooperation on low-carbon technologies and zero carbon breakthrough innovation is pursued in the absence of any emission reduction targets, the expected temperature increase is only slightly lower, at 3.5°C. The more environmentally aggressive policy architectures yield a temperature change of around 2.7°C, whereas intermediate efforts lead to a

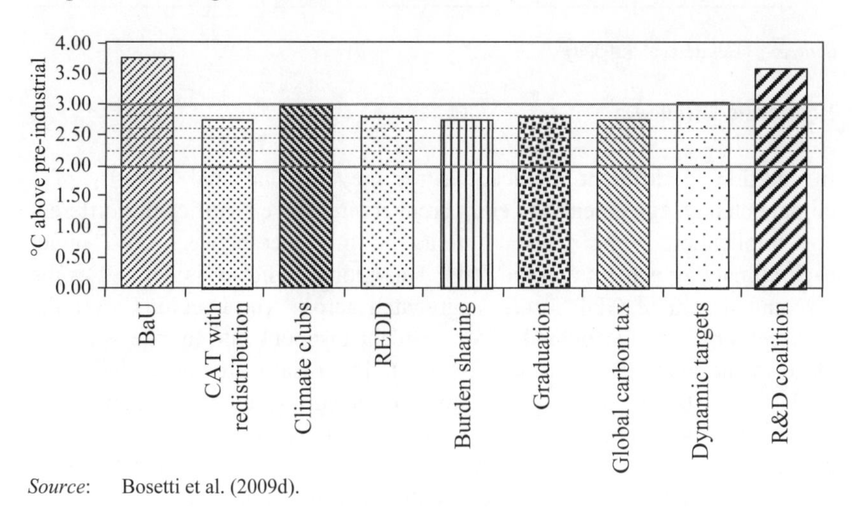

Source: Bosetti et al. (2009d).

Figure 3.12 Temperature change above pre-industrial levels in 2100

temperature increase of around 3°C. It is therefore clear that none of the policy architectures are able to keep temperature change below the 2°C threshold, as indicated in the Cancun Agreements.

In addition to the temperature change at the end of the century, it is likely that the pattern of temperature change would also have significant implications, in particular on ecosystems. It thus seems reasonable to prefer agreements that entail a smoother transition. Table 3.9 presents the number of 5-year periods for which temperature increase is greater than 0.1°C with respect to the previous 5-year period. It is clear that broad coalition entail fewer occurrences of temperature changes greater than 0.1°C, while the R&D coalition architecture, without an explicit emission target, is similar to the BaU scenario.

Table 3.9 Number of times that five-year temperature change is greater than 0.1°C

BaU	19
CAT with redistribution	12
Global carbon tax	12
REDD	12
Climate clubs	14
Burden sharing	12
Graduation	12
Dynamic targets	14
R&D coalition	19

Source: Bosetti et al. (2009d).

Economic Efficiency

In addition to different implications for the environmental integrity of a climate change agreement, the eight architectures have significantly different cost implications, approximated here as the difference between GWP under each climate agreement and the BaU. This global indicator is defined as the discounted sum of GDP losses, aggregated across world regions, over the next century, and discounted at a 5 percent discount rate (a rate which is close to the average market interest rate). The results are shown in Figure 3.13. The costs are directly proportional to the stringency of the stabilisation target: stabilising atmospheric CO_2 concentrations at 450ppm would cost between 1.2 percent and 1.49 percent of GWP. Climate Clubs and Dynamic Targets – which stabilise CO_2 concentrations at about 490 and 500 ppm respectively – entail moderate costs: around 0.32 percent and 0.24 percent of

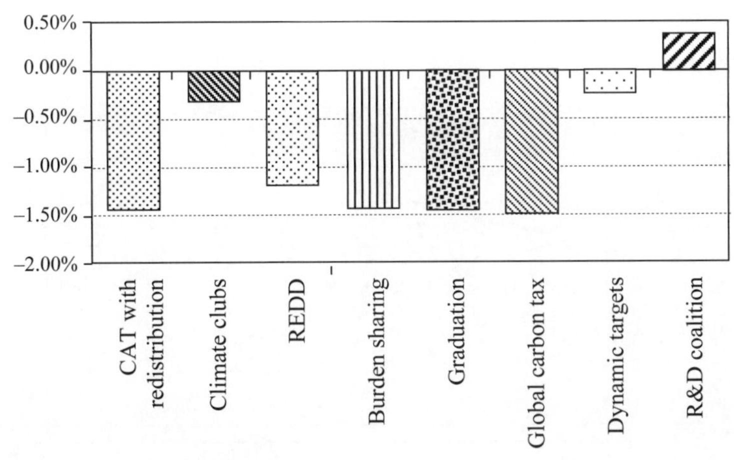

Source: Bosetti et al. (2009d).

Figure 3.13 Change in GWP with respect to BaU – Discounted at 5 percent

GWP respectively. Finally, the R&D Coalition leads to gains at the global level, of about 0.37 percent of GWP. These gains are explained by the positive effects of R&D cooperation, which reduce free-riding incentives on knowledge production.[25]

The analysis of how GDP changes over time for different world regions in the different scenarios may offer some indications as to what the distributional implications of climate change policies are, and it highlights winners and losers under each scenario. This information is likely to be important for policy makers and negotiators. When we look at the temporal distributions of the costs of the different architectures (see Figure 3.14) we see that the stringent architectures requiring universal and immediate action imply an immediate loss of GWP, rising up to 4 percent by the middle of the century. Gradual effort implies, on the other hand, less costly intervention at the beginning of the century. Only the global coalition based on R&D Cooperation leads to gains from 2040 – while implying short-term costs due to the diversion of resources to replenish the global R&D fund.

Equity and Distributional Impacts

The distribution of the costs and benefits of climate change and climate change policy is of paramount importance in determining both the feasibility and desirability of a specific architecture for agreement. Abstracting from the equity debate, the Gini Index for GDP in 2100 is used as a compact measure of distributional equity. The index represents the concentration of income among regions of the world, and shows inequality in income distribution (the

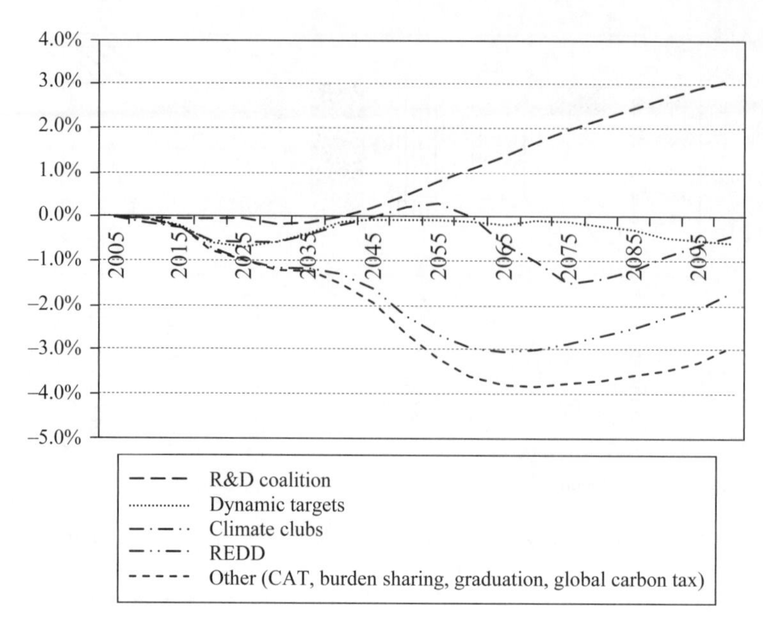

Source: Bosetti et al. (2009d).

Figure 3.14 Inter-temporal distribution of policy costs

lower the value of the indicator, the more equal the distribution of income). In all scenarios, there is an improvement in the distribution of income among world's regions with respect to the status quo. There are, however, differences across the climate change agreements (see Table 3.10): three policy architectures distribute the effort in a fair manner, according to income per capita and average per capita emissions, and thus lead to a more egalitarian distribution of income. When we consider immediate participation by all countries, we see that the carbon market can play a role in equalising the distribution of burden. Indeed, including REDD in the carbon market leads to an improvement in the distribution of income across regions, reflecting the fact that avoided deforestation is mostly an option in developing and tropical countries.

Enforceability and Feasibility

In the context of international agreements, enforceability and compliance are critical. Ideally, a global deal for climate change would sustain full participation and compliance, while ensuring an efficient level of emission reduction. Yet, because of the lack of a supranational institution able to enforce a climate policy, achieving a global agreement on GHG emission control may be very difficult if

Table 3.10 Gini index in 2100

BaU	0.2
CAT with redistribution	0.198
Global carbon tax	0.178
REDD	0.197
Climate clubs	0.158
Burden sharing	0.196
Graduation	0.158
Dynamic targets	0.156
R&D coalition	0.181

Source: Bosetti et al. (2009d).

not impossible (Barrett, 2003). By borrowing from game theoretic concepts, it is possible to derive quantitative measures of enforceability and political acceptability for all policies at both the global and regional levels.

We use the concept of potential internal stability (PIS, see Carraro et al., 2006) as a proxy for the theoretical enforceability of the agreement. This is a weak stability concept in the sense that an agreement is said to be potentially internally stable if the aggregate payoffs are at least as large as the sum of the regional payoffs in the BaU. If this condition is satisfied, all coalition members could be at least as well off as under the BaU scenario through a suitably designed transfer scheme. Global welfare is computed as the sum of welfare for each region. While the analysis is necessarily a simplification, the results do nonetheless provide a good starting point for the assessment of the enforceability dimension of climate agreements.

As shown in Table 3.11, all but one architecture imply an improvement in global welfare over the status quo – if one could design appropriate transfer schemes, then all regions could be made at least as well off as under the BaU scenario. The only exception is the autarkic coalition, where all countries, including developing countries, are required to undertake emission reductions domestically by imposing a carbon tax.

It is interesting to see that most of the welfare gains are experienced in developing countries: the stabilisation of the agreements would thus require the transfer of resources from developing to developed countries, which is unlikely to be politically acceptable.

While at the global level almost all architectures for agreement seem to be potentially enforceable, the picture is very different when we move down to the regional level, to explore the feasibility of the proposals. The last column of Table 3.11 shows the likely political feasibility of each policy architecture – approximated by the number of regions whose welfare would be higher

Table 3.11 Potential enforceability and political acceptability of the architectures

	Potential stability – World welfare (a)	Feasibility (b)
CAT with redistribution	0.68%	3
Global carbon tax	–0.17%	0
REDD	0.46%	4
Climate clubs	0.18%	6
Burden sharing	0.24%	3
Graduation	0.09%	3
Dynamic targets	0.20%	5
R&D coalition	0.10%	12

Notes: (a) Percentage of change with respect to BaU.
(b) Number of countries with +ve variation in welfare.

Source: Bosetti et al. (2009d).

than in the BaU scenario. Thus, the higher the number of countries that find a specific coalition profitable from an individual perspective, the more likely it is that the architecture is politically acceptable.

It is clear that in the R&D Coalition and Climate Clubs architectures – both involving some form of cooperation on R&D – a large share of countries find the agreement profitable (all and half of the countries are better off, respectively). The result on the climate club architecture is particularly interesting, as it seems to support the role of issue linkage in generating scope for gains from cooperation. The universal but incremental coalition based on Dynamic Targets is also likely to be politically feasible, as five out of 12 regions find it profitable: it is quite likely that a careful revision of the criteria for setting binding emission reduction targets could lead to a redistribution of welfare so that all countries would be better off.

3.6 CONCLUSIONS

From the analyses presented in this chapter, it is clear that achieving an environmentally meaningful and politically acceptable international agreement on climate change critically depends on the underlying economic incentives and disincentives that countries face individually.

Despite the necessary simplifications and assumptions in the analyses presented, a number of general lessons can be drawn that could prove useful in thinking about the future of a climate change deal.

First and foremost, a truly global action is needed if we are to successfully tackle the climate change challenge. The uncertainty surrounding climate change damages and mitigation costs, as well as the political uncertainty surrounding the achievement of a global deal, are stifling actions. Yet, the analyses presented in this chapter show that the most economically rational course of action would be to undertake some form of mitigation action immediately. This would lower the cost of compliance with any future targets, while at the same time leading to short-term benefits by internalising environmental and technology externalities.

Based on considerations of the Rio principles of common but differentiated responsibilities and respective capabilities, as well as an assessment of the political context, gradual participation of NA1 countries should be considered as a strategy to favour convergence of countries' positions on how to address climate change, though clear criteria and timelines are needed. Equity considerations and the overarching goals of poverty eradication should guide the definition of criteria and timelines for NA1 participation in a climate change mitigation regime.

NA1 countries should agree, in the immediate term, to limit their emissions to the BaU projection. On the one hand, this would allow them to participate in a global carbon market, thus lowering global costs of compliance and providing a tool for transferring resources from A1 to NA1 countries. On the other hand, incentives to free-ride would be minimised, and the environmental integrity of the agreement could be ensured. The inclusion of REDD could help both in providing incentives for developing countries to accept their BaU as their emission limit, and in lowering the policy costs.

It would also seem, however, that more decisive mitigation actions are needed if we are to achieve the target of limiting temperature change to 2°C above pre-industrial levels. The dilemma is present and clear: there is a trade-off between environmental effectiveness and economic efficiency, as well as between environmental effectiveness and political enforceability. Yet, our analyses indicate that a carefully designed system to transfer financial resources across regions could lead to an international agreement that is environmentally sound and politically feasible. This transfer mechanism could be built in the negotiated package, together with clear mitigation targets for A1 countries and a firm timeline and a transparent set of criteria to gradually bring NA1 into the system of binding emission reduction targets.

NOTES

1. This chapter is based on the following set of papers prepared using WITCH: Bosetti et al. (2013, 2009b, 2008c, 2009c, 2009a, 2009d).
2. According to current scientific knowledge, the stabilisation of atmospheric greenhouse gas

(GHG) concentrations at only 450 ppm CO_2 (roughly equivalent to 550 ppm all GHG included) is needed to keep global temperature change below dangerous levels, though even this target is thought unlikely to guarantee the stabilisation of temperature below 2°C.

3. A complete description of the model and a list of papers and applications are also available at http://www.witchmodel.org/.
4. While there are many other economic and political elements that can contribute in shaping countries' decisions (not) to engage in global cooperation, these are more difficult to capture in models of this type. This analysis therefore concentrates on the economic elements that factor in such a decision-making process.
5. The pure rate of time preference influences decision makers' perception over costs and benefits of mitigation action: the lower the pure rate of time preference, the more climate change damages are weighted in the decision process, as damages happen far in the future. In this analysis, two scenarios are considered, a high rate of time preference of 3 percent, and a low one, as advocated by Stern, of 0.1 percent.
6. Bosetti et al. (2013) discuss issues related to the weights attributed to welfare of regions in the coalitions.
7. The optimal abatement paths are estimated using the high-damage/low discount rate scenario, which is the most conducive to significant emissions reduction by the coalition and, as such, represents the best-case scenario for environmental effectiveness.
8. In this analysis, all countries are assumed to take action at the same time. An analysis of the implications of differentiated timing of action is provided in Bosetti et al. (2008c).
9. See Bosetti et al. (2007) for a description of how policy costs are computed in WITCH. In Bosetti et al. (2009b), damages from climate change are assumed to become infinite at the end of the century.
10. The results are robust to different probability distributions.
11. As pointed out in the introduction, new technologies that could achieve negative carbon emissions might allow a solution even in this case; however, this scenario has not been explored in this analysis.
12. For an analysis of emission trands and climate policy in China see Carraro and Massetti, (2012); for a comparison of China and India, see Massetti (2011).#
13. The results are robust across allocation schemes (equal per capita, sovereignty).
14. This analysis has appeared already in Bosetti et al. (2009d).
15. Most of the architectures considered in this paper are carefully described in Aldy and Stavins (2007).
16. This architecture is inspired by the work of McKibbin and Wilcoxen (2007).
17. Abatement cost curves for REDD were constructed using data from the Woods Hole Centre, and focus on Brazil.
18. This architecture is inspired by the work of Victor (2007).
19. Emission reduction effort for this group of countries is 5 percent with respect to BaU by 2020; 10 percent by 2030; and 30 percent by 2050.
20. This architecture is inspired by the work of Keeler and Thompson (2008) and Bosetti et al. (2008c).
21. This architecture is inspired by the work of Michaelowa (2007).
22. This architecture is inspired by the work of Frankel (2007).
23. This architecture is inspired by the work of Barrett (2007).
24. We use the MAGICC model to translate emissions and concentrations in temperature changes.
25. See Bosetti et al. (2008a) for a detailed analysis of knowledge spillovers.

REFERENCES

Aldy, J. and R.N. Stavins (eds) (2007), *Architectures for Agreement: Addressing Global Climate Change in the Post-Kyoto World*, Cambridge, UK: Cambridge University Press.

Barrett, S. (2003), *Environment and Statecraft*, Oxford: Oxford University Press.

Barrett, S. (2007), 'A multitrack climate treaty system', in J. Aldy and R.N. Stavins (eds), *Architectures for Agreement: Addressing Global Climate Change in the Post-Kyoto World*, Cambridge, UK: Cambridge University Press, pp. 237–259.

Bollen, J., B. Guay, S. Jamet and J. Corfee-Morlot (2009), 'Co-benefits of climate change mitigation policies: Literature review and new results', OECD Economics Department Working Papers No. 692, Paris: OECD.

Bosetti V., C. Carraro, M. Galeotti, E. Massetti and M. Tavoni (2006), 'WITCH: A World Induced Technical Change Hybrid Model', *The Energy Journal*, special issue on 'Hybrid modeling of energy- environment policies: Reconciling bottom-up and top-down', 13–38.

Bosetti, V., C. Carraro, E. Massetti and M. Tavoni (2007), 'Optimal energy investment and R&D strategies to stabilise greenhouse gas atmospheric concentrations', Nota di Lavoro 95.2007, Milan: Fondazione Eni Enrico Mattei.

Bosetti, V., C. Carraro, E. De Cian, R. Duval, E. Massetti and M. Tavoni (2009a), 'The incentives to participate in, and the stability of, international climate coalitions: A game theoretic analysis using the WITCH model', Nota di Lavoro 64.2009, Milan: Fondazione Eni Enrico Mattei.

Bosetti, V., C. Carraro, E. De Cian, E. Massetti and M. Tavoni (2013), 'Incentives and stability of international climate coalitions: An integrated assessment', *Energy Policy*, **55**(4), 44–56.

Bosetti, V., C. Carraro, E. Massetti and M. Tavoni (2008a), 'International technology spillovers and the economics of greenhouse gas atmospheric stabilization', *Energy Economics*, **30**(6), 2912–2929.

Bosetti, V., C. Carraro, A. Sgobbi and M. Tavoni (2009b), 'Delayed action and uncertain stabilization targets. How much will the delay cost?', *Climatic Change*, **96**, 299–312.

Bosetti, V., C. Carraro, A. Sgobbi and M. Tavoni (2009d), 'Modelling economic impacts of alternative international climate policy architectures: A quantitative and comparative assessment of architectures for agreement', in J. Aldy and R.N. Stavins (eds), *Post-Kyoto International Climate Policy: Implementing Architectures for Agreement*, Cambridge, UK: Cambridge University Press, pp. 161–172.

Bosetti, V., C. Carraro and M. Tavoni (2008c), 'Delayed participation of developing countries to climate agreements: Should action in the EU and US be postponed?', Nota di Lavoro 70.2008, Milan: Fondazione Eni Enrico Mattei.

Bosetti, V., C. Carraro and M. Tavoni (2009c), 'Climate policy after 2012', *CESifo Economic Studies*, **55**(2), 235–254, doi:10.1093/cesifo/ifp007.

Bosetti, V., A. Sgobbi, C. Carraro and M. Tavoni (2008b), 'Delayed action and uncertain target. How much will climate policy cost?', Nota di Lavoro 69.2008, Milan: Fondazione Eni Enrico Mattei.

Burniaux, J-M., J. Chateau, R. Dellink, R. Duval and S. Jamet (2008), 'The economics of climate change mitigation: How to build the necessary global action in a cost-effective manner', OECD Economics Department Working Papers No. 701, Paris: OECD.

Carraro, C., J. Eyckmans and M. Finus (2006), 'Optimal transfers and participation decisions in international environmental agreements', *The Review of International Organizations*, **1**(4), 379–396.

Carraro, C. and E. Massetti (2011), 'Editorial', *International Environmental Agreements, Law, Economics and Politics*, special issue on 'Reconciling domestic energy needs and global climate policy: Challenges and opportunities for China and India, **11**(3), 205–208.

Carraro, C. and E. Massetti (2012), ‘Energy and climate change in China’, *Environment and Development Economics*, **17**(6), 689–713.

Frenkel, J. (2007), ‘Formulas for quantitative emission targets’, in J. Aldy and R.N. Stavins (eds), *Architectures for Agreement: Addressing Global Climate Change in the Post-Kyoto World*, Cambridge, UK: Cambridge University Press, pp. 31–56.

International Energy Agency (IEA) (2002), *Beyond Kyoto: Energy dynamics and climate stabilization*, Paris: OECD/IEA.

Ingham, A., J. Ma and A. Ulph (2007), ‘Climate change, mitigation and adaptation with uncertainty and learning’, *Energy Policy*, **35**, 5354–5369.

IPCC (Intergovernmental Panel on Climate Change) (2007a), *Climate Change 2007: Synthesis Report*, Geneva: IPCC.

IPCC (Intergovernmental Panel on Climate Change) (2007b), *Climate Change 2007: Impacts, adaptation and vulnerability. Contribution of Working Group II to the Fourth Assessment Report on Climate Change* [M. Parry, O. Canziani, J. Palutikof, P. van der Linden, C. Hanson (eds)], Cambridge, UK and New York, NY, USA: Cambridge University Press.

Keeler, A. and A. Thompson (2008), ‘Rich country mitigation policy and resource transfers to poor countries’, paper presented at the Harvard Project on International Climate Agreements First Workshop, Cambridge, MA, 15–16 March 2008.

Kosobud, R.D.T., D. South and K. Quinn (1994), ‘Tradable cumulative CO_2 permits and global warming control’, *Energy Journal*, **15**, 213–232.

Massetti, E. (2011), ‘Carbon tax scenarios for China and India: Exploring politically feasible mitigation goals’, *International Environmental Agreements, Law, Economics and Politics*, special issue on ‘Reconciling domestic energy needs and global climate policy: Challenges and opportunities for China and India’, **11**(3), 209–227.

Massetti, E. and M. Tavoni (2011), ‘The cost of climate change mitigation policy in Eastern Europe, Caucasus and Central Asia’, *Climate Change Economics*, **2**(4), 341–370.

McKibbin, W.J., and P.J. Wilcoxen (2007), ‘A credible foundation for long-term international cooperation on climate change’, in J. Aldy and R.N. Stavins (eds), *Architectures for Agreement: Addressing Global Climate Change in the Post-Kyoto World*, Cambridge, UK: Cambridge University Press, pp. 185–208.

Michaelowa, A. (2007), ‘Graduation and deepening’, in J. Aldy and R.N. Stavins (eds), *Architectures for Agreement: Addressing Global Climate Change in the Post-Kyoto World*, Cambridge, UK: Cambridge University Press, pp. 81–104.

Richels, R. and J.A. Edmonds (1995), ‘The economics of stabilizing atmospheric CO_2 concentrations’, *Energy Policy*, **23**(4/5), 373–378.

Victor, D. (2007), ‘Fragmented carbon markets and reluctant nations: implications for the design of effective architectures’, in J. Aldy and R.N. Stavins (eds), *Architectures for Agreement: Addressing Global Climate Change in the Post-Kyoto World*, Cambridge, UK: Cambridge University Press, pp. 133–160.

Weyant, J.P. and J. Hill (1999), ‘The costs of the Kyoto Protocol: A multi-model evaluation, introduction and overview’, *The Energy Journal,* special Issue on ‘The costs of the Kyoto Protocol: A multi-model evaluation’, 1–24.

Wigley, T.M.L., R. Richels and J.A. Edmonds (1996), ‘Economic and environmental choices in the stabilization of atmospheric CO_2 concentrations’, *Nature*, **379**, 240–243.

4. Coping with Uncertainty

Massimo Tavoni

Economic analysis of climate change has emerged as a major area of research, with important repercussions for policy advice.[1] Within this topic, integrated assessment modelling is a quintessential instrument because it allows to characterise most of the features of climate change, and to inform climate policy. Among the many subjects that have been addressed through modelling, the issue of uncertainty emerges as one of the most prominent. Climate change is known to be characterised by a cascade of uncertainties that cover every aspect of the problem: from climate change science, to climate change impacts and finally to climate change solutions. The recognition of these issues is so important that the IPCC remarked the need for a 'risk management approach' when dealing with climate change. Two of the mostly widely discussed papers in the past few years (Stern, 2006, and Weitzman, 2009) have brought uncertainty right at the heart of their analysis, and have shown the extent to which it affects the solution space. The next IPCC assessment report in its third working group has dedicated one chapter to uncertainty. Even in informal discussions among informed readers of climate change, uncertainty is likely to be one of the topics of convergence, and potentially, of consensus.

Although the methodological tool for incorporating uncertainty into climate change modelling is known, its implementation has often been limited, since it requires interdisciplinary competences and normally leads to considerable complications for both model formulation and solution. In this chapter, we report some of the analysis under uncertainty that has been carried out at FEEM in the course of recent years, which has involved the modification and extension of the WITCH model to be able to include the most salient features of uncertainty.

In Section 4.1 we assess the role that uncertainty plays in the design of climate policy instruments: we use Monte Carlo to assess which among a carbon tax or a cap-and-trade system is best suited to deal with climate mitigation, showing the advantages of carbon taxation but also emphasising the role of trading in promoting innovation. Then, in Section 4.2, we move to the much-debated issue of the timing of mitigation action, and whether in the face of uncertainty, it is optimal to anticipate or postpone emission abatement. Since our stochastic formulation of the WITCH model provides support for early, moderate action as a hedging tool against uncertainty, we explore the issue of how the mitigation strategies should be affected by uncertainty. We do this in Section 4.3, and examine the role of uncertainty of innovation, which allows us to derive the conditions under which innovation uncertainty leads to more R&D investments.

Generally, our analysis indicates that it is important to include uncertainty when dealing with the economic analysis of climate change, and that in many circumstances uncertainty exacerbates the results of the deterministic analysis. Our contribution is also useful for indicating which methodological tools can be best, or most easily, handled when dealing with the numerical modelling of climate change economics.

4.1 POLICY INSTRUMENT CHOICE UNDER UNCERTAINTY

We begin the analysis by studying how uncertain abatement costs and uncertain climate sensitivity affect the design of policy instrument, either via a carbon tax or cap-and-trade. This is an important problem that has led to important analytical and numerical work (Weitzman, 1974; Pizer 1999; Newell and Pizer, 2003), and had fostered a debate on price versus quantity instruments, which has continued for decades. While the presence of uncertain abatement costs pushes risk adverse individuals to prefer the price instrument, the randomness of climate damages introduces an opposite bias towards the quantity instrument. We study how these two competing forces combine and we comment on the resulting optimal policy choices for a risk adverse individual.

To this end we use the WITCH model to simulate a climate stabilisation policy of 450 parts per million (ppm) of CO_2 under two different policies:[2]

Cap-and-trade. We use as policy tool for a world carbon market that equalises marginal abatement costs worldwide. Each model region is assigned an entitlement in terms of emissions rights that it can either use or trade within the market. The total amount of emission allowances produces a global emission path over time, which entails the stabilisation of concentrations below the target.

Global carbon tax. Each model region receives an entitlement of emissions rights. The total amount of emission rights produces a global emission path over time that entails the stabilisation of concentration below the target. Emissions above and below the regional level of emission rights are taxed or subsidised, respectively. The carbon tax is set equal to the price of emission permits, for instance to the price of carbon, derived from the cap-and-trade scenario.

We portray uncertainty on two main variables: abatement costs and climate sensitivity. Regarding abatement costs, given the focus of the WITCH model on energy R&D, uncertainty has been modelled as concerning the effectiveness of energy knowledge investments. In particular, we have concentrated on the parameter accounting for the productivity of investments in energy R&D. It is very difficult to find an adequate probability distribution estimate in the literature for anything related to uncertainty about abatement costs. We have assumed that the productivity parameter is multiplied by a log-normal variable with unit mean and 0.3 standard deviations. This ensures that the productivity of new R&D investments varies by roughly 50 percent for a 95 percent confidence interval.

As for climate sensitivity, we have fitted the distribution of the climate sensitivity parameter on data available from Hegerl et al. (2006). The distribution of the parameter Climate Sensitivity is assumed to be logN (0.96, 0.485), see Figure 4.1. Such a distribution corresponds to a mean climate sensitivity of 2.9, the base value normally used as input into the WITCH model.

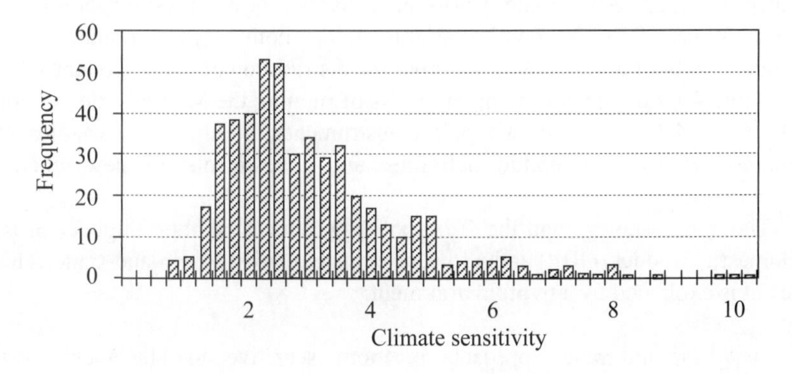

Figure 4.1 Probability density function for the climate sensitivity

Results

The present exercise aims to determine the magnitudes and distributions of the optimal choice variables (investments and consumption) in response to random realisations of the energy R&D productivity parameter and the parameter that governs climate sensitivity under the cap-and-trade and under the carbon tax scenarios. We simulate the case of an advisor who uses the central value for the productivity parameter to advise a policy maker. The policy maker has the option to choose between cap-and-trade and a carbon tax (fixed once and forever) as a policy instrument. If the real world were deterministic, the Weitzman principle of equality between the two instruments would be valid (Weitzman, 1974). Suppose, instead, that the policy maker knows that the world she faces is characterised by uncertainty on the realisation of the productivity parameter and of the climate sensitivity. If she decides to adopt a cap-and-trade approach to contain carbon emissions, she has a flexible instrument that will automatically adjust carbon prices to match equalised marginal abatement costs under any random realisation of the R&D productivity parameter. Policy costs might turn out to be higher, but the time path of emissions remains unchanged. The presence of uncertainty on climate sensitivity does not neutralise the possibility to miss the environmental target. However, given the stringency of the climate policy, deviations from the optimal temperature path typically occur in a range in which the effect on the climate damage function is contained. If instead she implements the carbon tax, she will not be able to adjust the carbon price to match the random realisation of marginal abatement costs. Thus policy costs will not exceed a maximum threshold, but the environmental target might not be met, even with fully deterministic climate dynamics. The randomness of the climate system further increases the variability of the final temperature outcome. With our Monte Carlo analysis we provide estimates of the distributions of key economic variables under both types of policies, as a function only of our assumptions about the distribution of R&D productivity.

Table 4.1 summarises the main results of running the Monte Carlo version of the model under the two policy instruments, on the key variables of interest such as economic activities and investments in new energy technologies.

The table shows that the Tax policy produces slightly higher Gross Domestic Product (GDP) and a lower variance relative to cap-and-trade. This result is explained by a twofold argument:

1. The cap-and-trade approach is more sensitive to abatement cost realisations: whenever R&D is found to be less productive than anticipated, economic policy costs, in terms of lower GDP and consumption, are higher with the quantity instrument than with the price instrument.
2. Second, environmental penalty arises when, under the tax policy instrument, emissions are higher than expected due to high abatement costs, which does not counterbalance greater policy costs. Even if there might be combinations of high abatement costs and high climate sensitivity. This derives from the stock nature of the CO_2 externality (temperature and damages react very slowly to increase in emissions flows) and from the limited impact of variations of the world temperature when a stringent environmental target is implemented.

Table 4.1 Results of the Monte Carlo analysis

		Cap and trade	Carbon tax
GDP	Mean	2500.9	2501.7
	Standard Dev.	16.2	13.7
Energy R&D	Mean	1170.6	1226.2
	Standard Dev.	336.2	333.3
Wind+Solar	Mean	857.5	855.9
	Standard Dev.	57.8	44.6

Considering the impacts of the two different policies on innovation activities, such as R&D investments, we find that the distribution of investments in energy efficiency R&D is higher under the tax policy, while the variance is higher under the cap-and-trade policy. Differences in the productivity of R&D have asymmetric effects under the carbon tax and cap-and-trade policies. The higher variance of the investments in R&D under cap-and-trade is due to the fact that with a quantity instrument, the objective is, quite trivially, to meet the cap. This requirement binds when energy R&D productivity is low (some investment has to be made anyway) but also when R&D productivity is high (enough investment has to be done to meet the target). This is the reason for the higher variance. Concerning the mean, there is no incentive with a quantity instrument to invest beyond what has to be done to meet the cap. However, such an incentive does exist when considering a price instrument.

Finally, when looking at renewables, we find that the distribution of investments in wind-and-solar electricity generation technologies is higher under the cap-and-trade policy. Also the variance is higher under the cap-and-trade policy compared to the tax policy instrument.

Discussion

This section has analysed the impact of uncertainty in abatement costs and in climate sensitivity for quantity versus price policy instruments. For this purpose we have used the integrated assessment WITCH model to perform Monte Carlo analyses on the realisation of the energy R&D productivity parameter and the sensitivity parameter, which governs temperature increases. This is just one way in which abatement cost uncertainty might be represented in this model. Results indicate that uncertainty about abatement costs leads to GDP and consumption profiles with slightly higher means and considerably lower variance under the price instrument (carbon tax). Emissions are constant under the cap-and-trade scenario, while they adjust in response to random abatement costs under the carbon tax scenario (not necessarily satisfying the limits required by a stabilisation target). The effect of uncertainty on energy R&D productivity produces higher values for energy R&D investments under the price instrument.

The presence of uncertainty on climate sensitivity does not revert the standard result shown in Pizer (1999). This is due to the fact that at low levels, temperature increases, as in the case when a stringent climate target is adopted, deviations from the optimal path are relatively mild when a quadratic damage function is adopted.

Implementing either a global trading scheme or a tax scheme could involve huge transfers of resources between countries. We have shown how large these transfers can be in the case where the initial allocations were made on a per capita basis. It is reasonable to question the practicality of such large transfers. In this context we note that a different allocation of initial allocations can reduce the size of the transfers. Moreover it is also important to note that the size of the transfers is not a function of the choice of taxes or permits. In both cases a given initial allocation implies the same magnitude of transfers. The major difference is that with a tax a central authority is needed to act as a 'clearing house' for the transfers, whereas with permits such transfers can be made in a more decentralised manner.

4.2 THE TIMING OF MITIGATION ACTION UNDER AN UNCERTAIN FUTURE

Despite the growing concern about actual on-going climate change, there is little consensus about the scale and timing of actions needed to stabilise the concentrations of GHG. Many countries are unwilling to implement effective mitigation strategies, at least in the short-term, and no agreement on an ambitious global stabilisation target has yet been reached. It is thus likely that some, if not all countries, will delay the adoption of effective climate policies. This delay will affect the cost of future policy measures that will be required to abate an even larger amount of emissions. Also, given the uncertainty about the scale and timing of future climate targets, it is important to analyse whether the expectation of gaining better information should induce policymakers to wait before acting. Or should the possibility that impacts of climate change might turn out much worse than we now believe lead policymakers to take more stringent immediate action to avoid abrupt and costly policy changes in the future?

We address both questions by quantifying the economic implications of delayed mitigation action, and by computing the optimal abatement strategy in the presence of uncertainty about a global stabilisation target (which will be agreed upon in future climate negotiations).

The Cost of Delaying Action

We begin by exploring the global implications of different decisions about the course of action to be followed in the coming 20 years on the Net Present Value (NPV) of policy costs for this century. In particular, we assess the sensitivity of 450/550 ppm CO_2 only policy costs to a 20-year delay in action, and also the cost of initiating a climate mitigation policy (either stringent or mild) and subsequently dropping it. Results are shown in Table 4.2.

Table 4.2 The cost implications of delayed action, NPV GWP loss to 2100, discounted at 3 percent (and 5 percent)

	Continue along the Business-as-Usual path	Undertake a mildclimate control policy (550 ppm)	Undertake a stringent climate control policy (450 ppm)
Take action now	.	0.3% (0.2%)	3.5% (2.3%)
Wait 20 years on the Business-as-Usual path	.	0.4% (0.3%)	7.6% (5.5%)
Wait 20 years on a mild policy path	−0.03% (0.06%)	.	4.2% (2.7%)

If world policymakers jointly start taking action now to control climate change, the net present value of stabilisation costs at a 3 percent (5 percent) discount rate ranges from 0.2 percent (0.3 percent) to 3.5 percent (2.3 percent) of Gross World Product (GWP), depending on whether we undertake a mild policy (550 ppm CO_2 only, equivalent to 650 ppm all gases) or a stringent policy (450 ppm CO_2 only, equivalent to approximately 550 ppm all gases). These figures represent the additional cost of stabilisation, compared to a world in which no costly abatement actions (the BaU scenario) are undertaken.

As shown in Table 4.2, the cost of delaying actions for 20 years varies considerably. In particular, it crucially depends on whether policymakers decide to undertake mild or stringent policy action after the delay period. While the cost of delaying to undertake mild actions are relatively modest (the net present value of the cost is equivalent to 0.4 percent of GWP as opposed to 0.2 percent of GWP with an immediate start), moving from the Business-as-Usual (BaU) to a stringent climate stabilisation target after the 20-year delay is extremely costly; up to 7.6 percent (5.5 percent at 5 percent discounting) of the net present value of GWP over this century. This represents either an increase of policy costs of about 130-140 percent, or an equivalent loss of 5.7 (2.2 at 5 percent discounting) trillion USD per year of delay. Of course, this is likely to disrupt the common understanding of the economic feasibility of any stringent action as currently perceived by policymakers.[3] Another important result emerges from our analysis: a policy strategy that immediately begins to undertake some emissions reductions that are consistent with a 550 ppm CO_2 only stabilisation target, and in 20 years reverts to the BaU scenario, does not harm global welfare, and actually leads to a very mild increase of GWP, thanks to the internalisation of the various externalities on carbon, exhaustible resources, and innovation. The same mild mitigation policy can be tightened at lower costs than continuing along a BaU path for the next 20 years: the net present value of the costs of shifting from a mild to a stringent policy reaches 4.2 percent (2.7 percent) of GWP, still well below the cost of inaction for 20 years followed by embracing a stringent climate policy in 20 year's time.

The policy implications of this exercise are quite clear, and support the arguments that call for immediate action to tackle climate change (see Stern, 2007, 2008). If we continue doing nothing for 20 years, the costs of shifting from a BaU to a stringent climate policy are extremely high. On the other hand, undertaking some form of mild stabilisation policy seems to be a hedging strategy which, at virtually no cost, would allow us to revert to BaU and, at relatively modest cost, to undertake more decisive action if a more stringent stabilisation policy is decided upon.

4.3 THE TIMING OF MITIGATION UNDER CLIMATE TARGET UNCERTAINTY

In this section, we use a stochastic programming version of the WITCH model to investigate an optimal policy strategy given the uncertainty about the stringency of the stabilisation target that will eventually be established in an international climate agreement. We frame the analysis on a scenario tree, solve for all scenarios simultaneously and account for non-anticipativity constraints (the action has to be the same for different scenarios before the disclosure of uncertainty, while the optimal reaction to the information revealed when uncertainty is eliminated is allowed afterwards).

This formulation enables us to devise the optimal strategy before uncertainty is disclosed, and to identify potentially optimal hedging behaviour. It also enables us to determine the most suitable portfolio of mitigation technologies given the uncertainty on stabilisation targets that will be adopted in the future. Let us assume that uncertainty about the climate target is resolved in 2035. However, today, the future target is unknown and three scenarios are assumed to emerge with equal probability:

1. *BaU (no target)*: no constraint on emissions. No agreement on a common target is achieved, and countries are free to adopt their own BaU emission levels. This is an extreme scenario in which there is a complete stall in international negotiations and consequently single regions decide not to commit to any mitigation from 2035 onward. It could also be read as a scenario in which by 2035 either we discover that climate change theories are unfounded (or un-anthropogenic) or we discover a perfect geo-engineering solution that allows carbon emissions to be decoupled from climate warming.
2. *550 ppm CO_2 stabilisation.* We assume an international climate agreement is reached, with a target of carbon concentrations in 2100 of 550 CO_2 only (650 all GHG).
3. *450 ppm CO_2 stabilisation.* We assume an international climate agreement is reached, with a target of carbon concentrations in 2100 of 450 CO_2 only (550 all GHG).

We adopt a uniform probability distribution to express an uninformative a priori that foresees equal chances of any of the three scenarios materialising.

Using the above assumptions, we compute the optimal investment path for all energy technologies, for physical capital and for R&D as a non-cooperative equilibrium of the game among the 12 world regions represented in WITCH. Figure 4.2 reports the resulting global carbon emissions trajectories over this century for the stochastic case and the deterministic cases of no target, 550 and 450.

The figure shows the three different deterministic cases vis à vis the stochastic one that takes into account the fact that all three are possible. It provides clear evidence of hedging behaviour: the optimal strategy before uncertainty is resolved to engage in significant mitigation. In 2030, emissions are 57 percent lower than in the BaU scenario, essentially the same as they are in 2008. That is to say, in a world that has an equal chance of being confronted with no climate policy, a mild policy or a stringent policy in 30 year's time, the best strategy in the short-term (for the next two decades) is to minimise emission growth. Emissions in 2030 would also be 23 percent lower than the target prescribed by the mild policy, and only 14 percent higher than the more stringent one.

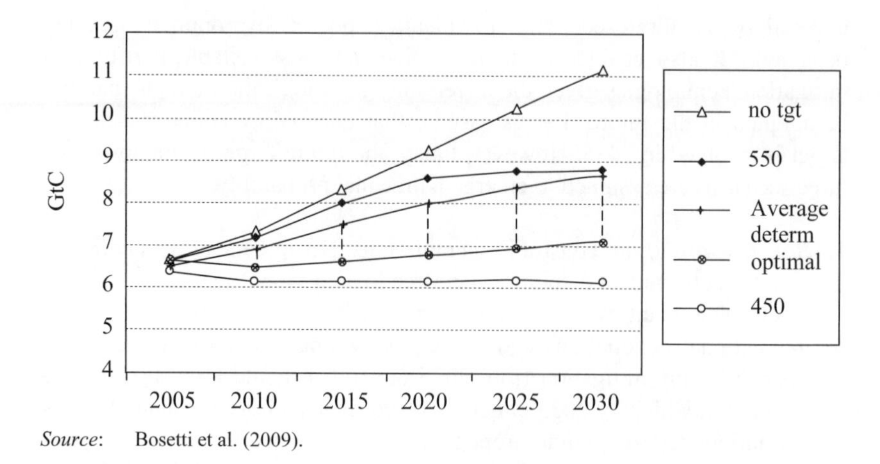

Source: Bosetti et al. (2009).

Figure 4.2 Fossil fuel emissions for optimal stochastic and deterministic case, to 2030

The low level of emissions in the stochastic scenario is driven by the possibility of a stringent climate target, albeit this has just a 33 percent chance of occurring. The convexity of marginal abatement costs in the mitigation levels is the reason for this risk-averse approach. As shown in Figure 4.2, the optimal emission path is always below the average of the three deterministic cases, by 6 Gt CO_2 in 2030, or 22 percent. This quantifies the dimension of the precautionary strategy.

The Economic Cost of Sub-optimal Strategies

Another interesting exercise is to quantify the economic loss resulting from non-compliance with the hedging strategy detailed in the previous section, and the pursuit of the optimal investment strategies for the deterministic cases. In order to do this, we run the stochastic version of the model, but fix all the choice variables to those of the three deterministic cases until 2035 (the time frame to uncertainty resolution). From that point on, the model is free to optimise for each of the three states of the world, given the sub-optimal choices undertaken during the first periods. This enables us to quantify the economic losses that would result from sticking to either a BaU Scenario, a 550 ppm stabilisation policy or a 450 ppm stabilisation scenario (thus not adhering to the optimal strategy) in the periods before the final policy outcome is known.

The first important result is that following the BaU scenario from now to 2035 does not allow a feasible solution to the optimisation problem if the target after 2035 is the ambitious one (450 CO_2 (550eq)). In other words, a

BaU strategy for the next three decades precludes the attainment of the more stringent stabilisation target. This result extends the one seen in the first part of the chapter, which showed that policy costs increase sharply with the period of inaction. It should be noted that accounting for technologies that can achieve negative carbon emissions might allow a solution even in this case; however, such a scenario is highly speculative and has not been included in this analysis.

Sub-optimal strategies aiming at 550 and 450 ppm result in more and less emissions (respectively) than the optimal strategy does before 2035 – and vice versa thereafter – as noted in the previous section. We thus expect that following a mild 550 ppm policy would lead to improved outputs in the first decades, followed by deterioration, and that the opposite would be true if we followed an emission mitigation policy that was more stringent than the optimal one, such as aiming for the 450 target. This is what is shown in Figure 4.3, where the gap between the expected value of GWP for the sub-optimal cases and the optimal one is shown.

One can see that committing to more mitigation results initially in economic loss, but leads to a higher output after the resolution of uncertainty, due to the fact that the burden of meeting the stringent target in the future has been alleviated. On the other hand, a sub-optimal 450 strategy inflicts costs that are somewhat higher than the benefits of a milder choice (–0.8 percent GWP losses in 2035 for 450, as opposed to +0.45 percent gains for 550) in the short-term. The picture reverses after 2035, when the costs of under-

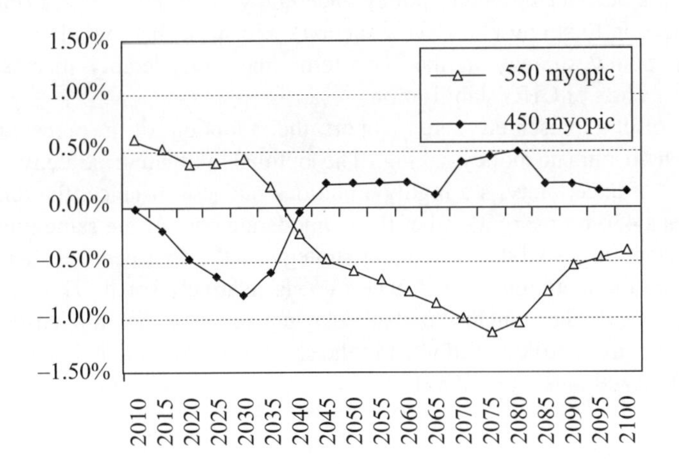

Source: Bosetti et al. (2009).

Figure 4.3 GWP differences of sub-optimal strategies wrt the optimal hedging one

abatement before shifting to a stringent policy are higher than the benefit of a more virtuous early strategy.

Discussion

This section constitutes an attempt to provide better information to policymakers by identifying the short-term implications of uncertainty about future climate targets and by computing the cost of delaying the introduction of effective mitigation strategies.

Despite general warnings on the risks of inaction, most estimates of the cost of mitigation policies have tended to underestimate the cost of delaying action. In the first part of the section, we show that this cost is far from negligible, even in a model in which technology is flexible and endogenous, and R&D investments are optimally chosen. We quantify the cost of a 20-year delay in action as an increase of GWP losses of about 140 percent, or in the range (depending on discounting) of 2.2–5.7 trillion USD per year of delay. We also show how committing to a mild abatement effort in the short-term might substantially reduces the cost of delaying action.

In the second part, we analyse the short-term policy implications of the uncertainty surrounding future climate targets. Results clearly point to precautionary behaviour in which emissions are considerably reduced below the expected value of the deterministic cases. Finally, we quantify the economic inefficiencies of sub-optimal policy strategies (for instance those that do not account for future policy uncertainty) in different cases (ranging from inaction to strong emission reduction). We show that the adoption of a 'wrong' policy strategy in the short-term may considerably increase the economic costs of GHG stabilisation.

The results presented here support the adoption of a precautionary approach to climate policymaking. The optimal mitigation strategy in the presence of uncertainty is a highly ambitious one (for instance, the one that achieves a 450 ppm target) rather than a moderate one. At the same time, the cost of adopting a moderate climate strategy in the first decades, and then shifting to the ambitious one if necessary, is relatively small. This result is consistent with the conclusion that delaying action is not too costly if a moderate climate policy is adopted in the short-term (but it may become very costly if no action is taken at all).

4.4 UNCERTAINTY AND INNOVATION

Technological change is an uncertain phenomenon. In its most thriving form, ground-breaking innovation is so unpredictable that any attempt to model the uncertain processes that govern it is close to impossible. Despite the complexities, research dealing with long-term processes, such as climate change, would largely benefit from incorporating the uncertainty of technological advance. Yet, bringing uncertainty into models has proved particularly difficult, especially with regards to technological change.

On a more general level, the challenge of modelling endogenous technological change in all its features, including randomness, becomes increasingly important when dealing with the analysis of stringent climate targets. Many energy-economy models have been used to perform cost effectiveness of climate policies. Not surprisingly, the related literature has produced a dispersed range of costs estimates for these policies, resting on the different formulations and assumptions that stand behind each model. The recognition of the relevance of this issue has led researchers to model technological change as an endogenous process, although typically in a deterministic fashion. The existing literature accounting for uncertainty has mostly concentrated on the uncertainty affecting climate damages and abatement costs, as well as other parameters, such as the discount factor. Few studies have looked at the consequences of uncertainty (say of climate change impacts) on innovation efforts. However, little focus has been devoted to the analysis of the intrinsic uncertainty of innovation, and how this uncertainty might change results and policy recommendations.

This section delves into the issue of uncertain technological progress when a climate obligation is in place. In particular, we seek to analyse different optimal responses in terms of investments and climate policy costs when we model innovation as a backstop technology characterised by either a deterministic or an uncertain process.

A Simple Model of Uncertain Innovation

To this scope, we first develop a simple analytical model. We use a two-period, two-technology model where the social planner minimises costs but needs to achieve a given environmental target. Given a target level of abatement to be undertaken during the second period, the planner can choose a combination of two carbon-free technologies: a traditional technology (say nuclear fission) and an advanced, backstop technology (say nuclear fusion). Abatement costs with the backstop technology are initially higher than with the traditional one, but can be reduced by investing in R&D during the first period. We introduce uncertainty by modelling the R&D outcome on the abatement cost of the backstop technology as uncertain: the innovation effort leads to a central value reduction in abatement costs with a given probability p, and to lower and higher abatement costs states with probability (1-p)/2, respectively. The high cost state represents the failure of the R&D program:

abatement costs are not reduced by the innovation effort, and remain higher than the traditional carbon-free technology costs for any level of abatement. In this case, the planner chooses not to operate the backstop technology, because it is too costly, and resorts to the cheaper, traditional technology. The low cost state represents a greater than expected success of the R&D program: backstop technology costs are always lower than in the central case. The lower the cost is, the higher the abatement pursued with the advanced technology.

Solving the analytical model yields two main insights:

- While the abatement cost using the backstop technology in the central case are equal to the average of the low and high R&D effectiveness cases, the total cost of meeting the environmental target are higher for the central certain case.
- We assume that marginal benefits of innovation increase with abatement using the backstop technology (through learning by doing). Then, for interior solutions for the abatement variables, investments in innovation increase with uncertainty.

How does this finding translate into real life considerations? First, one has to bear in mind that the social planner can pick from a variety of technologies to achieve an environmental target, say, to reduce CO_2 emissions. The process of innovation is uncertain and investing in R&D implies the risk of failing. However, if R&D is less productive than expected existing technologies will still allow meeting the mitigation target, though at a higher costs. If R&D is successful, on the other hand, the benefits would be higher than they would have been in the central case. This payoff asymmetry is such that the upside of super productive innovation outweighs the downside of failure. Hence, in the presence of innovative technologies, a risk-neutral planner would choose to invest more when the R&D outcome is uncertain.

Numerical Analysis

In order to investigate the role of uncertain technological change, we focus on the carbon-free backstop technology. Innovation can lower the price of this otherwise non-competitive technology, but it is modelled in a stochastic setting to account for the uncertainty of the R&D outcome. A climate target of 450 ppm is assumed throughout the analysis. We compare the deterministic case with the uncertain formulation. The average of the latter coincides with the deterministic one to ensure the equivalence of the comparison exercise. In the uncertain formulation there is a 50 percent chance to achieve the central case and a 25 percent chance to achieve the failure and best cases, respectively.

Since we are investigating the role of uncertainty on innovation, it is interesting to compare the R&D investments in the stochastic case and in the equivalent deterministic case, before uncertainty is resolved in 2050. Results of investments on innovation are presented in Figure 4.4; the graph shows that optimal R&D investments are always higher in the stochastic formulation with respect to the deterministic case before the resolution of uncertainty. The numerical analysis thus confirms that modelling R&D as having an uncertain outcome induces more innovation effort, as predicted by the analytical example outlined above. As expected, in the stochastic setting, once uncertainty is resolved, R&D is higher for the best case than for the central, and it is zero for the failure state.

In order to test the results for robustness and to understand the effect of key assumptions, we have repeated simulations for a different set of assumptions on entry time and the level of risk aversion. In Figure 4.5 we present the R&D results when we assume different entry times of the backstop technology ('early' in 2040, and 'late' in 2060). The picture shows that early resolution of uncertainty on the efficacy of the R&D programme leads to a higher level of optimal R&D investments. The contrary holds in the case of late discovery of the program's effectiveness. Although the effect on the levels of investment is significant, entry time has a small impact on policy costs.

As a concluding analysis, we drop the assumption of risk neutrality and investigate what happens when the central planner is risk-averse. In this case, lower utility is attached to risky investments, and thus we expect to find an effect contrary to the results presented so far. We start by analysing the unit risk

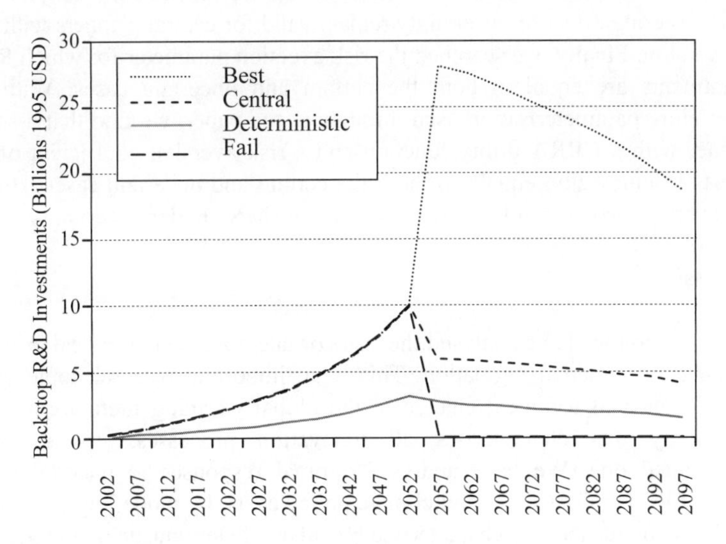

Figure 4.4 R&D investments for backstop

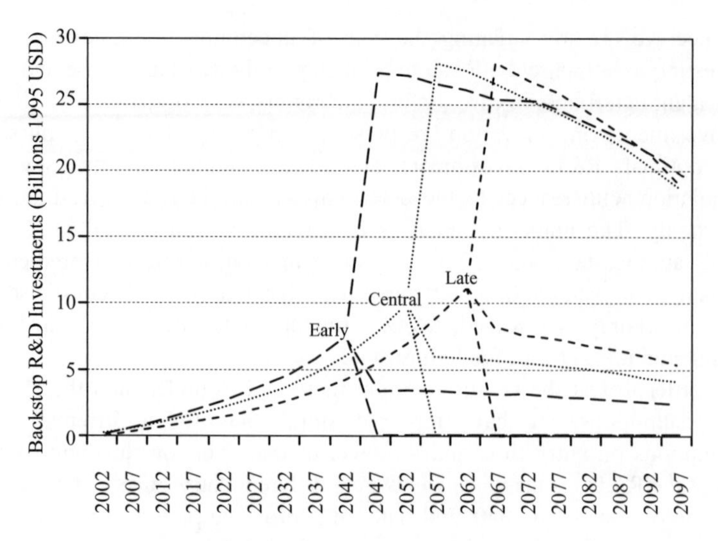

Figure 4.5 Effect of entry time on backstop R&D investment

aversion case of logarithmic utility function. Numerical results show that R&D investments in the uncertainty case are indeed lower than for the reference risk neutral analysis. The risk aversion increase roughly halves innovation effort: for example, R&D investments in 2050 drop from ten to five USD billions. Despite this effect, they remain higher than for the certain case (that for example has 2.2 USD billions investments in 2050), thus confirming that the R&D fostering effect of uncertainty remain valid for central planners with unit risk version. Finally, we searched the risk aversion parameter for which R&D investments are equal in both the certain and uncertain cases. With the uncertainty parameterisation used throughout the paper, we find that a social planner with a CRRA utility function and a risk aversion coefficient of 1.5 invests in innovation equally in both the certain and uncertain cases. Higher risk aversion would result in lower innovation shares under uncertainty.

Discussion

In this section we have analysed the issue of uncertain technological progress within environmental regulation. This is an important research topic given the relevance of technical change in the global warming literature and the uncertainty that characterises all innovation processes, yet a poorly investigated one. We have analysed optimal responses to uncertainty, in terms of R&D investments and climate policy costs, by modelling innovation as a backstop technology characterised by either a deterministic or an uncertain process. For this purpose, we have developed a simple analytical model and modified the hybrid integrated assessment model WITCH to account for a carbon-free backstop technology dependent on uncertain R&D realisations. We have performed a stochastic cost effectiveness analysis of a CO_2 stabilisation policy of 450 ppm.

Numerical results, in accordance with analytical insights, have shown how modelling innovation in a backstop technology as an uncertain process leads to higher optimal levels of R&D investments. A detailed representation of the energy sector has allowed us to capture path dependency in technological evolution, and therefore to account for the consequences of different innovation efforts on technology deployment and externality resolution. To check for the robustness of the results, we have shown how different timings of backstop availability affect R&D investments and policy costs in the expected direction but to a limited extent in terms of magnitude. Finally, the role of social planner risk aversion has been analysed and has shown to have a counterbalancing effect that reduces the gap in innovation investments with and without uncertainty.

4.5 CONCLUSIONS

This chapter has investigated the role of uncertainty for climate change control policies. We have begun by highlighting the major role that uncertainty plays in the case of climate change, which is characterised by pervasive risks and uncertainties. Its investigation cannot thus be considered complete unless it incorporates uncertainty as one of the fundamental tools of analysis. In this chapter we have used the specifically developed version of the WITCH model that has been devised to account for uncertainty.

We have shown that uncertainty plays an important role when choosing climate policy instruments, such as a carbon tax versus a cap-and-trade system. Uncertainty also considerably affects the timing of mitigation action, and our stochastic analysis has provided at least partial support for a precautionary principle with gradual ramp up of abatement as the optimal hedging strategy. We have also shown that the uncertainty related to innovation of clean energy technologies is an important determinant of the most efficient R&D strategy.

Although the application of uncertainty and risk analysis to climate change is still in its infancy, this chapter has shown that the tools of integrated assessment models can be successfully enhanced to account for at least the more salient aspect of climate change uncertainty. The various stochastic applications of WITCH collected in this chapter can serve as a useful starting point for the important work waiting ahead of us.

NOTES

1. This chapter is drawn or partly reproduces the following articles: Bosetti et al. (2008, 2009) and Bosetti and Tavoni (2009).
2. If not otherwise stated, concentration targets in this chapter refer only to carbon dioxide.
3. As mentioned in chapter 1 and discussed further in chapter 7, the possibility to absorb GHG from the atmosphere by means of biomass with carbon capture and storage (BECCS) or using direct air capture methods would reduce the cost of delaying mitigation action.

REFERENCES

Bosetti, V., A. Golub, A. Markandya, E. Massetti and M. Tavoni (2008), 'Abatement cost uncertainty and policy instrument selection. A dynamic analysis', in A. Golub and A. Markandya (eds) (2008), *Modelling Environment-Improving Technological Innovations under Uncertainty*, Oxon, UK and New York, USA: Routledge, pp. 127–157.

Bosetti, V., C. Carraro, A. Sgobbi and M. Tavoni (2009), 'Delayed action and uncertain targets. How much will climate policy cost?', *Climatic Change*, **96**(3), 299–312.

Bosetti, V. and M. Tavoni (2009), 'Uncertain R&D, backstop technology and GHGs stabilisation', *Energy Economics*, **31**, S18–S26.

Hegerl, G.C., T.J. Crowley, W.T. Hyde and D.J. Frame (2006), 'Climate sensitivity constrained by temperature reconstructions over the past seven centuries', *Nature*, **440**(20), April: 1029–1032.

Newell, R.G. and W.A. Pizer (2003), 'Regulating stock externalities under uncertainty', *Journal of Environmental Economics and Management*, **45**(2), 416–432.

Pizer, W. (1999), 'Optimal choice of policy instrument and stringency under uncertainty: The case of climate change', *Resource and Energy Economics*, **21**, 255–287.

Stern, N. (ed.) (2006), 'The economics of climate change: The Stern review', London: London HM Treasury.

Stern, N. (2007), *The Economics of Climate Change: The Stern Review*, Cambridge: Cambridge University Press.

Stern, N. (2008), 'Key elements of a global deal on climate change', London: The London School of Economics and Political Science.

Weitzman, M. (1974), 'Prices vs. quantities', *The Review of Economic Studies*, **41**(4), 477–491.

Weitzman, M. (2009), 'On modelling and interpreting the economics of catastrophic climate change', *Review of Economics and Statistics*, **91**(1), 1–19.

5. Climate Policy and the Forestry Sector: The Role of Non-energy Emissions

Valentina Bosetti

While it is broadly agreed that policies to reduce emissions from deforestation will be critical for fighting climate change, the linkage of international forestry and other land sectors to compliance markets for greenhouse gas (GHG) emissions reductions remains a critical policy issue.[1] Policies for Reducing Emissions from tropical Deforestation and Forest Degradation (REDD) offer the opportunity to mitigate a major share of global GHG emissions at low estimated costs based on existing technologies. Investments in REDD are also a potentially attractive 'wooden bridge' for reducing near-term emissions while buying time to reengineer other sectors of the economy (Chomitz, 2006). A key policy question is how to balance funding of low-cost emission reductions from tropical forest conservation and other land-based activities that are available in the near-term with investments to drive technological innovation in energy, transport, and industrial sectors over a longer time period.

While the framework for limiting greenhouse gases under the Kyoto Protocol excluded mechanisms to reduce deforestation, there is a growing consensus on including REDD as a critical element of a future global climate policy regime. The Copenhagen Accord of December 2009 calls for immediately establishing a mechanism to finance REDD and other forestry sequestration activities in developing countries (UNFCCC, 2009). The Accord specifically calls for exploring both public and private market-based financing approaches, but the details remain to be determined. Governments and other organisations have put forth multiple proposals for financing REDD activities, including market-based approaches with different degrees of fungibility between forest carbon credits and GHG reductions in other countries and sectors. Policymakers in the United States are also considering multiple means of financing international forest carbon activities within emerging regional compliance markets for GHG reductions as well as in recent legislative proposals for a cap-and-trade system at the Federal level.

The 2010 Cancun Agreements recognise that REDD+ is not only about reducing emissions but also about halting and reversing forest loss, thus emphasising the need to ensure that the existing forest should be maintained. The Agreements also recognise that all actors should play a role in reducing human pressures on forests, thus giving part of the responsibility in the hands of those countries and actors that create demands which drive deforestation.

Under a carbon-market system, mitigation of tropical forest emissions, perhaps measured at a national scale against a reference level of historic emissions, would generate credits that could be sold and traded in a market for GHG emissions permits or 'allowances' that could be used to satisfy legally-binding emissions control obligations. Such an approach offers the potential to channel resources to the most cost-effective mitigation opportunities, lowering the costs of climate policies and generating significant financing for REDD over the near-term (Murray et al., 2009). Furthermore, REDD has the potential to channel significant financial resources to developing countries, thus improving the prospect of a climate change agreement.

One set of concerns is that linking international forest carbon credits to GHG compliance markets could lower near-term costs at the expense of reductions in developed countries and associated incentives to develop critical low-carbon technologies. Concerns over the potential of REDD credits to 'flood' compliance markets and dampen innovation have been largely voiced with regards to the scale of potential forest carbon credits relative to the size of the European Union's existing Emissions Trading Scheme (EU ETS) market. For example, the European Commission cited a potential 'imbalance' between the supply and demand for REDD credits as one of the reasons for its recommendation to defer the inclusion of REDD in the EU ETS at the end of last year (EC, 2008). In this chapter we discuss the analyses published in Tavoni et al. (2007) and Bosetti et al. (2011), based on the linkage of the WITCH model with some of the top world forestry models, that investigate the implications of linking REDD credits to a global carbon-market. We focus on the effect that including REDD in the carbon-market has on policy cost, on the carbon-market, and on the consequences for technology innovation in the energy sector. We also evaluate the implications for policy flexibility when future emission reduction targets are uncertain and an unexpected tightening of climate policy might be needed in coming decades as a response, for example, to new scientific information about climate change impacts. The first analysis (Tavoni et al., 2007) adopts a soft link approach[2] between the WITCH model and Brent Sohngen's forestry model (described in Sohngen et al., 1999), and investigates the effect of REDD on a moderate climate policy. The second study (Bosetti et al., 2011) builds upon the first and, by means of a hard link[3] of the WITCH model with three forestry models, examines the interaction between REDD and a more stringent climate policy.

The first analysis emphasises how, in the context of a moderate climate policy, including REDD as a mitigation option is economically efficient, but might lead to dynamic inefficiencies. Indeed, the availability of alternative mitigation technologies, that might be delayed by the availability of REDD, could be as important as a hedging strategy in the event, for example, that new scientific information requires a sudden downward revision of climate targets in the future. The second analysis concludes that concerns over REDD discouraging technological innovation are largely misplaced, if the global abatement effort, even in the short-run, is significant. Reducing the costs of climate change protection by steering efforts into the lowest marginal cost options for mitigation is precisely the economic rationale for an emission trading system, providing a net gain for the whole society as long as the right long-term emission reduction targets are in place. Furthermore, allowing the banking of emission credits generated through REDD helps cushion the risk of higher costs when there is less than perfect anticipation of increases in future emission-reduction targets. Of course, if there is a concern that forest carbon credits will be too plentiful, policy makers always have the option of limiting the numbers of credits allowed in the system (for example, through a cap on total REDD credits). It is surprising that the EU has not taken this up in its revised ETS proposals.

In general, we can conclude that we should not lose sight of the costs of excluding REDD from the carbon-market: doing so risks making climate change protection policies unnecessarily expensive and misses important opportunities to enable political agreement on more stringent GHG reduction targets now and in the future.

5.1 PREVIOUS STUDIES OF THE IMPACTS OF GLOBAL FORESTS ON CLIMATE STABILISATION, CARBON MARKETS, AND TECHNOLOGICAL INNOVATION

A growing literature analyses the potential role of reductions in deforestation and other land-based mitigation activities as part of global climate change policies. Researchers have estimated that forest-sector emission reductions, largely in the tropics, would contribute half as much abatement as the total energy sector under an economically optimal strategy that balances the costs versus the benefits of avoiding climate change (Sohngen and Mendelsohn, 2003). Previous studies using integrated economy-climate models have focused on the potential contribution of forestry and other land-based activities to a least-cost portfolio of mitigation options to achieve a particular target level of GHG concentrations. In particular, results from the Energy Modelling Forum 21 at Stanford University and related efforts indicate that reducing deforestation, as well as afforestation/reforestation, changes in forest management, and agricultural activities to sequester carbon and reduce emissions, can play a significant role in meeting stabilisation targets and reducing the costs of climate policy over the next century (Rose et al., 2008; Fisher et al., 2007).

Apart from differences in modelling details, the estimated savings from REDD critically depend on the policy target for GHG concentrations in the atmosphere and on the menu of options for reducing emissions that are available under each policy scenario. More mitigation alternatives bring more potential sources of low-cost emission reductions, reducing the reliance on any single source of cost savings. Another critical assumption affecting the estimated role of REDD across models is the expected development of future bioenergy technologies. In particular, biomass production for electricity generation combined with carbon capture and sequestration (BECCS) could be a powerful competitor for land allocated to forest-based carbon sequestration, if it became a feasible means to generate energy with negative carbon emissions (Obersteiner et al., 2001). At the same time, studies have also shown that policies to reduce emissions from fossil fuels could create perverse incentives that increase emissions from deforestation as bioenergy crops expand, unless parallel incentives are in place to avoid emissions from land use (Wise et al., 2009; Edmonds, 2003).

To the extent that policies to reduce forestry emissions lower the cost of climate protection, REDD may enable greater global emission reductions than could be achieved without REDD for the same overall policy cost.

Integrated assessments of the contribution of the forest sector to global climate stabilisation policies (Rose et al., 2008) have, in general, abstracted from the institutional details of how national commitments would be structured and how emission reductions would be traded in a carbon-market system. Studies have focused on the economically efficient pattern of mitigation, assuming that all countries immediately start to follow the optimal trajectory to stabilise emissions. A more realistic pattern of participation across regions would result in an economically sub-optimal global path of emissions. In particular, delayed participation by developing countries and other regions misses low-cost mitigation opportunities in the near-term and causes international leakage (shifts in emissions) that increase the need for more costly future emission reductions to achieve climate targets (Calvin et al., 2009).

More recently, studies using partial equilibrium models have examined the impacts of including forest carbon credits within a carbon-market given different scenarios of negotiated national-level emission reduction commitments and potential restrictions on trading. All these studies find that including forestry mitigation significantly lowers policy costs. The specific results depend on the assumptions about the supply of forest carbon credits as well as the demand, which is determined by the reduction targets, the trading restrictions, and available alternatives for mitigation in other sectors. For example, in an update of Anger and Sathaye (2008), Dixon et al. (2009) find that compliance costs are reduced by one third as a result of the introduction of credits from forestation and reduced deforestation into a single-period market ending in 2020, based on announced targets from Annex 1 nations. They estimate that the inclusion of REDD credits lowers the carbon price by 45 percent in the case of no restrictions on REDD credit purchases and by 20 percent when REDD credits are restricted to 20 percent of Annex 1 mitigation. Based on a similar static framework, the Eliasch (2008) report for the UK Office of Climate Change estimates the costs of reducing global emissions to 50 percent of 1990 levels by 2050 (475 CO_2 equivalent stabilisation) may be lowered by 25–50 percent in 2030 and 20–40 percent in 2050 when both reduced deforestation and forestation credits are included. They also estimate that forestry credits from developing countries would lower the European Union's carbon price in 2020 by 4 percent to 41 percent, depending on whether the EU commits to 20 percent or 30 percent reductions below 1990 levels by 2020. These impacts decline with assumed 'supplementarity' limits on the share of emission-reduction requirements that can be satisfied through REDD and other international credits.

Recent studies have examined the potential carbon-market impacts of REDD in a dynamic framework that includes the possibility of credit 'banking'. When long-term targets are sufficiently ambitious and anticipated, regulated entities have a potential incentive to over-comply with current requirements and 'bank' excess allowances or other types of credits for use in later periods when allowance prices are higher, as it is likely in the case of tightening commitments to reduce emissions (Dinan and Orszag, 2008; Murray et al., 2009). When banking is active, the estimated path of carbon prices is generally higher in the near-term and lower in the long-term compared to a scenario without the flexibility to trade emission reductions over time. The path of climate variables also reflects a higher initial effort given banking, revealing the suboptimal timing of abatement when the path of reductions is simply fixed to a realistic set of reduction commitments, without considering the profit-maximising timing of activities by market participants.

Taking banking into account, assuming national commitments consistent with a stabilisation target of about 550 ppm CO_2 equivalent, and using marginal abatement cost curves from the Global Timber Model (GTM), Piris-Cabezas and Keohane (2008) and Murray et al. (2009) indicate that a program to reduce tropical deforestation emissions would lower the global carbon price by 13 percent and 22 percent, respectively from 2013 to 2050.

Also including credits from afforestation/reforestation and changes in the management of timber plantations would reduce the price by a total of 31 percent and 43 percent, respectively, over this period. Similarly including banking and using recent forestry cost estimates from GTM, a dynamic analysis with the ADAGE and NEMS models reports similar results on REDD (USCAP 2009).

REDD credits could affect carbon price volatility as well as price levels. Piris-Cabezas and Keohane (2008) argue that a reservoir of banked REDD credits could provide firms with a buffer against unexpected price spikes and volatility in the future. In addition to banking, option contracts based on available sources of low-cost abatement, such as REDD, could provide an alternative form of flexibility to help hedge against carbon-market volatility. Golub et al. (2008) examine the potential market for REDD-backed call options that a buyer could exercise at a future date to purchase REDD credits at an agreed upon price, depending on the carbon prices prevailing at that time. Such transactions could provide an alternative source of near-term financing for REDD programs while allowing carbon-market participants to hedge firm-level risks associated with permit price spikes. Moreover, such options could buy time and flexibility for the development of new energy technologies such as carbon capture and storage (Golub et al., 2009).

5.2 KEY FEATURES OF THE ANALYSES WITH THE WITCH MODEL

The use of the WITCH model provides three advantages compared with previous analyses of the impacts of including deforestation reductions and other forest sector activities in a global carbon-market. First, WITCH enables us to treat technological change endogenously. Consideration of induced technical change is critical for assessing the value of forestry mitigation as a 'bridge' to facilitate the transition to a future low-carbon economy. We model how REDD alters incentives through the carbon-market and how these, in turn, affect technology innovation and deployment in the energy sector. In contrast, previous studies of linking forest-based credits to carbon-markets take future technologies and abatement costs as fixed.

Second, our study is intertemporal and dynamic, which is essential for modelling how technological change evolves endogenously. The dynamic model also allows us to explore the impact of a market framework in which participants can 'bank' credits in anticipation of more ambitious emissions cuts in the future. This likely feature of carbon-markets has significant implications for the timing and pattern of abatement across forestry and other sectors.

Third, WITCH is an integrated assessment model linking the economy and the climate. Rather than only examining the market impacts, the link to the climate allows consideration of how the predicted patterns of mitigation affect GHG concentrations and the associated climatic implications. This enables explicit analysis of the costs to meet stabilisation targets, as well as the degree to which cost savings from forest carbon mitigation can enable more ambitious stabilisation objectives. The climate module also allows us to examine alternative estimates for the costs of reducing emissions from deforestation while accounting for differences in the projected Business-as-Usual (BaU) levels of future emissions from global forests. Accounting for varying trajectories of forest emissions is essential for evaluating the role of global forests in meeting stabilisation targets. We also use the integrated assessment framework to examine the role of reductions in deforestation under different scenarios on how stabilisation policies might evolve over time.

The remainder of the chapter is divided into three sections. Sections 5.3 and 5.4 describes how we augmented the WITCH model to include the REDD option and the alternative sources of supplemental data on the marginal costs of reducing tropical deforestation. In Section 5.5, we report and discuss the results, focusing on the impacts on deforestation, policy costs, carbon prices, technological change, and financial flows among countries. Section 5.6 concludes this chapter.

5.3 THE WITCH MODEL AND THE FORESTRY ESTIMATES

Most integrated assessment models such as WITCH do not directly examine avoided deforestation and other land-use mitigation activities but they can be linked to forestry and land-use models. We present results using both 'soft' and 'hard' link approaches to examine the potential of land-based activities in climate stabilisation.

Soft-link Approach

In Tavoni et al. (2007), WITCH is coupled with the forestry model that builds upon the one described in Sohngen et al. (1999), which is also used by Sohngen and Mendelsohn (2003) to analyse global sequestration potential. The forestry model used in this analysis contains an expanded set of timber types, as described in Sohngen and Mendelsohn (2006). There are 146 distinct timber types in 13 regions: each of the 146 timber types modelled can be allocated into one of three general types of forest stocks. First, moderately valued forests, managed in optimal rotations, are located primarily in temperate regions. Second, high-value timber plantations are managed intensively. Subtropical plantations are grown in the southern United States (loblolly pine plantations), South America, southern Africa, the Iberian Peninsula, Indonesia, and Oceania (Australia and New Zealand). Finally, low-valued forests, managed lightly if at all, are primarily located in inaccessible regions of the boreal and tropical forests. The inaccessible forests are harvested only when timber prices exceed marginal access costs. The forestry model maximises the net present value of net welfare in the forestry sector. One important component of the costs of producing timber and carbon are land rental costs. The model accounts for these costs by incorporating a series of land rental functions for each timber type. The rental functions account for land competition between forestry and agriculture, although they are not presently responsive to price changes in agriculture (see Sohngen and Mendelsohn (2006) for additional discussion of the land rental functions). Incentives for carbon sequestration are incorporated into the forestry model through the rental value of carbon. The price of energy abatement is the value of sequestering and holding a ton of carbon permanently. The rental value for holding a ton of carbon for a year is determined as the path of current and future rental values on that ton that is consistent with the price of energy abatement currently. One of the benefits of using the rental concept for carbon sequestration is that the carbon temporarily stored can be paid while it is stored, with no payments accruing when it is no longer stored (for instance, if forest land is converted to agriculture, or if timber is harvested, the forest is left in a temporarily low-carbon state). Furthermore, renting carbon does not penalise current forestland owners by charging them for emissions. We do, however, account for long-term storage of carbon in wood products by paying the price of carbon for tons when they are stored permanently after harvest. For simplicity in this analysis, we assume that 30 percent of harvested wood is stored permanently, following Winjum et al. (1998).

This first approach makes the most of the information and characteristics of both models. However, given their complexities, they are integrated via an iterative procedure. In order to do so, both models are augmented so that they can incorporate results from the other, and are run for subsequent iterations until they convergence, as measured by a sufficiently small rate of variation of carbon prices. We define this as being less than a 5 percent average deviation in prices and quantities from one scenario to the next. To make the two models consistent, several additional adjustments were made. First, the different regions had to be matched. Coincidentally, the regional disaggregation is similar in the two cases – 12 regions for the WITCH model and 13 for the forestry one – so that only minor adjustments were needed.

Also, the WITCH model has 5-year time steps and the forestry model has 10-year time steps. To link the two, we utilised prices at the 10-year intervals provided by the WITCH model in the forestry model. We interpolated carbon sequestration rates between 10-year time increments from the forestry model when incorporating forest sequestration in the WITCH model. The forestry model has been augmented to comprise the time path of carbon prices, which is equalised across regions and given by the emissions permits prices of the cap and trade policy. To account for the non-permanence of the biological sequestration, carbon prices are transformed into annual storing values via rental rates. For more information, see Sohngen and Mendelsohn (2003). The energy-economy-climate model has been fed the carbon quantities sequestered by forests in each region by counting them in the carbon emission balances, as well as in the budget constraint – at the carbon price value. As expected, the initial high responses of both models – in terms of adjustments of carbon prices to the quantities sequestered in forests and vice versa – gradually shrink, and an equilibrium is achieved after 11 iterations. The average deviation is 3 percent for prices, whereas for quantities it is 4 percent.

5.4 HARD-LINK APPROACH

The second analysis is performed by supplementing the WITCH model with cost curves for reducing tropical deforestation estimated by different forestry and land-use models. We augment the WITCH model with estimates generated by separate land-use and forestry models under carbon price scenarios that are consistent with the WITCH model predictions.

While the soft-link approach presented above iterated between WITCH and a forestry model to increase consistency, here the advantage is that we can draw on the results that a broader set of forestry and land-use models, by including alternative supply cost curves. This 'hard-link' approach, where the forestry model is substituted with a reduced form model, as it does not imply a complete integration of models nor the iterative process between models, has the advantage of being computationally much faster. The disadvantage is that it does not capture potentially significant feedbacks between the energy and forestry sectors as well as with other economic sectors and the overall climate system. Integration of forestry and other land uses within economy-wide models is an on-going area of climate policy research (see Hertel et al., 2009, for an overview), including consideration of trade effects (Golub et al., 2009) and feedbacks between climate change and agricultural productivity (Ronneberger et al., 2009). Our approach uses WITCH to focus on the energy sector and technology innovation impacts and then explores the sensitivity of results to alternative estimates from the separate forestry and land-use models.

Given uncertainty surrounding both tropical forestry emission baselines and the costs of reducing deforestation for different regions and time periods, we consider three different sets of REDD cost curves. Each set of curves is derived from a distinct modelling framework external to WITCH. We focus on avoided deforestation (the first 'D' of REDD) for which a broad range of cost estimates is available, as this is the focus of policy discussions. By abstracting from institutional details, transaction costs, and other real-world complications, these models estimate the theoretical potential of reducing deforestation emissions, but likely understate the true economic costs of reducing deforestation emissions in practice. Greater availability of REDD as a means to capture carbon allows to better detect its potential impacts in terms of suppression of key technologies, such as carbon capture and storage, often deemed strategically important in climate change policy discussions.

One set of supply curves comprises the estimated compensation needed to cover 30 years of opportunity costs of reducing deforestation emissions in the Brazilian Amazon based on modelling from the Woods Hole Research Centre (Nepstad et al., 2007). This analysis integrates spatially explicit partial-equilibrium models of potential land rents from soy, timber, and cattle ranching along with a simulation model of future deforestation considering expected road development and protected area scenarios. While only one out of many potential participating countries in a global REDD program, Brazil accounts for up to one half of global deforestation in the humid tropics (Hansen et al., 2008), and has the most developed current infrastructure for monitoring and implementing REDD. In August 2008, the government of Brazil established a fund to protect the Amazon forest with the goal of raising USD21 billion in international contributions over the next 13 years. In December 2008, Brazil further announced a voluntary commitment to reducing its deforestation by 70 percent over 2006 to 2017 relative to the average deforestation levels over the previous decade (Government of Brazil 2008). Given Brazil's institutional capabilities, expressed commitment to reducing deforestation, and market-developments supporting low-deforestation agriculture, Brazil-only REDD is potentially a realistic scenario for near-term REDD potential (see Nepstad et al., 2009 for details). Nevertheless, the Woods Hole Research Centre's modelling does not account for potential 'leakage', shifts in deforestation to other parts of the world, as a consequence of deforestation reductions in Brazil's Amazon region. As a result, these estimates will likely understate the cost of achieving global emissions reductions by means of a REDD program limited to the Brazilian Amazon.

We also consider two estimates of the global potential for reducing deforestation emissions, based on a scenario in which all tropical forest nations immediately join a carbon trading system and have the institutional and governance capacity to fully implement deforestation-reduction programs.

In reality, countries vary widely in both their willingness and ability to reduce deforestation. The global model estimates simulate an idealised forest carbon policy in which any changes in the modelled stocks of forest carbon relative to the BaU are continually rewarded or penalised at the carbon-market price. In practice, a global REDD system could provide a less comprehensive set of incentives, thereby discouraging participation by some nations. Fewer sources of mitigation and greater potential for international leakage would undermine the effectiveness of reductions achieved in any particular location and further increase the costs of reducing deforestation emissions. The global models thus estimate the maximum economic potential for reducing deforestation emissions under an optimal REDD system.

The global REDD estimates used for this study are from two of the leading economic models of global forests, based on scenarios for rising carbon prices consistent with those in our policy simulations. We consider results from the Global Timber Model prepared for the Energy Model Forum 21 (GTM EMF21) at Stanford University (Sohngen and Sedjo, 2006). GTM is a dynamic partial-equilibrium model that optimises changes in deforestation, afforestation, and forest management across ten world regions, accounting for the competition between forests and agricultural uses. This is the same model as the one used in the previously described soft-link analysis (Tavoni et al., 2007). It is also used by the US Environmental Protection Agency for its analyses of international forest carbon provisions in proposed US climate legislation. We use the GTM EMF-21 estimates based on carbon prices rising at a real rate of 5 percent annually. This scenario approximates the exponential growth of carbon prices predicted by WITCH under the scenarios with banking. However, this scenario will tend to overestimate REDD supply in the cases without banking, which result in a lower near-term and higher long-term price. Although GTM predicts changes in forest management, afforestation, and biomass supply for bioenergy as a result of climate policies, we only include the estimates for reduced deforestation emissions.

As an alternative, we also incorporate estimated costs of reducing deforestation emissions from linked runs of the IIASA (International Institute for Applied Systems Analysis) model cluster (Gusti et al., 2008), prepared for the UK Office of Climate Change as part of the Eliasch Review (Eliasch, 2008). These estimates are based on a spatially-explicit global model of global forestry and agricultural land use (The Global Forestry Model; G4M) linked to a partial equilibrium model (Global Biomass Optimisation Model; GLOBIOM) that accounts for feedbacks between land allocation and the prices of land-based commodities. We use IIASA's estimates based on carbon prices rising at a cubic rate, as this set of runs was the most consistent with our policy scenarios. In contrast to the GTM results, the cubic price path used by IIASA closely approximates the carbon price trajectory predicted by

WITCH under our scenarios without banking, while relatively underestimating REDD potential in the scenarios with banking, which result in higher near-term and lower long-term carbon prices. While the IIASA model cluster also estimates afforestation/reforestation and forest biomass for bioenergy, we only incorporate the results for reduced deforestation.

These models also differ in the BaU levels of forestry emissions as well as in the estimated costs and quantities of available reductions. The varying BaU arise from differences in the underlying data on land-use and carbon, assumptions over deforestation drivers, and regional coverage (see Kindermann et al., 2008, for a comparison of the global models). We account for these differences by adjusting the BaU land-use emissions in WITCH to be consistent with each REDD scenario. Due to the differing structures and variables in each model, we do not attempt to reconcile any of the socioeconomic assumptions in WITCH with those underlying the different deforestation scenarios. This means that there is only partial consistency in our modelling of the forestry and non-forestry sectors.

The models provide a wide range of estimated REDD potential. The IIASA analysis estimates the greatest absolute potential emission reductions from reducing deforestation. At a price of USD11, USD23, and USD55 per ton of CO_2, IIASA estimates global deforestation emission reductions of 1.8 billion, 2.6 billion, and 3.1 billion tons of CO_2 in 2010 and 3.0, 3.7, and 3.9 billion in 2030. The estimated reductions from these estimates from GTM are about 25 to 50 percent lower in 2010 and 50 to 65 percent lower in 2030. This is driven largely by lower projections of baseline emissions from deforestation. At a price of USD11, USD23, and USD55 per ton of CO_2, the estimated reductions from this set of GTM estimates are 0.8 billion, 1.9 billion and 1.9 billion tons in 2010 and 0.7 billion, 1.0 billion, 1.6 billion tons in 2030.

Although limited to the Brazilian Amazon, the WHRC modelling estimates significant potential reductions from deforestation as the estimated opportunity costs of avoiding deforestation in the Brazilian Amazon are lower than those from the global models. One reason for the lower costs is that the global models incorporate price feedbacks, with avoided deforestation efforts raising the global market price of agricultural land, thus making reductions in deforestation more costly. Another reason is that WHRC considers timber revenues from sustainable forest management as a benefit from conserving forests, in contrast to the GTM and IIASA models where timber harvests are a potential benefit of land conversion. The WHRC estimates assume constant baseline emissions from deforestation in the Brazilian Amazon of 916 million tons of CO_2 per year and estimates that 570 reduced for less than USD11 in 2010, with costs rising sharply thereafter. In 2030, WHRC estimates reductions of 760 and 800 million tons for opportunity costs less than USD11 and USD24 per ton, respectively.

Finally, we note some additional limitations of the study. We do not consider other potentially significant forest sector mitigation opportunities from afforestation/reforestation and changes in forestry plantation management, which could be significant in both Annex 1 and tropical and non-tropical developing countries. For simplicity, we do not consider potential interactions between the forest and energy sectors, other than those mediated through the carbon-market. We do not consider the potentially important impacts of climate policy for bioenergy demand and the consequences for deforestation. Potential feedbacks between the climate and forests, which could affect both emissions and mitigation potential, were also beyond the scope of this study.

5.5 MODEL RESULTS

Soft-link Analysis

In this section we report the numerical results of the first analysis (Tavoni et al., 2007), focusing on the contribution of forestry management in meeting a CO_2 (only) stabilisation policy of 550 ppm by 2100. To give the feeling of what such a policy entails in terms of global warming mitigation, in Figure 5.1 we show the time profile of carbon emissions for a BaU and a 550 ppm policy resulting from using the WITCH with abatement only in the energy sector. In a no-policy scenario emissions grow to 73 Gt CO_2 by the end of the century, whereas for the 550 ppm policy, emissions peak around 2050, falling by more than half after that with respect to the BaU. The 550 ppm policy decreases the carbon intensity in the economy considerably, and reduces the increase in global temperature by 2100 to 2.2°C, from 2.9°C in the BaU. Although this temperature is still higher than 2°C, the level advocated by the IPCC, we concentrate on this target given its relevance, especially in terms of political feasibility.

We start by reporting the potential of forestry in contributing to the foreseen emission reductions, and then analyse the impacts on the carbon-markets and the policy costs. Finally, we examine the implications for the energy abatement portfolio, with a particular look at the implications for induced technological change.

Figure 5.2 reports carbon abatement over the century accomplished through forestry in OECD and non-OECD countries *vis à vis* the overall abatement effort. The picture underlines an important role for biological sequestration: forests sequester around 275 Gt CO_2 cumulative to 2050. This estimate is consistent with the results presented in earlier IPCC reports but of course there are costs associated with this forestry effort. Overall, forestry contributes to 1/3 of total abatement to 2050, or three wedges in the words of

Pacala and Socolow (2004). After the peak in emissions in 2050, the share of forestry in total abatement starts to decline (from 2050 to 2100 it increases by only 10 percent in absolute values), given that the target becomes more stringent and permanent emission cuts in the energy sector are needed.

The largest share of carbon sequestration occurs in non-OECD countries during the early part of the century (Table 5.1). Around 63 percent of all of

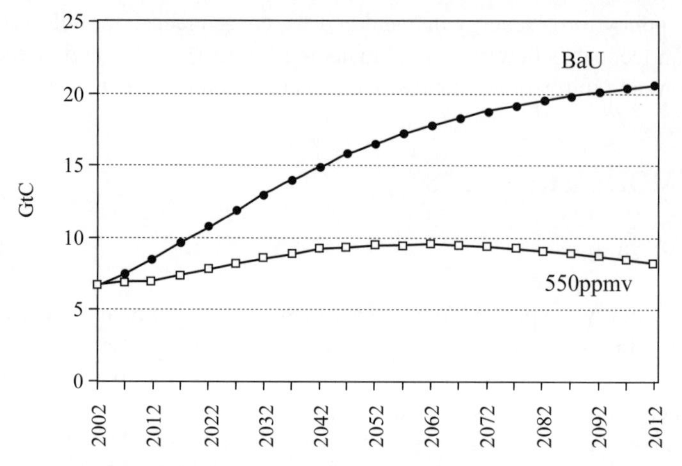

Source: Tavoni et al. (2007).

Figure 5.1 Carbon emissions for Business-as-Usual and 550ppm policy

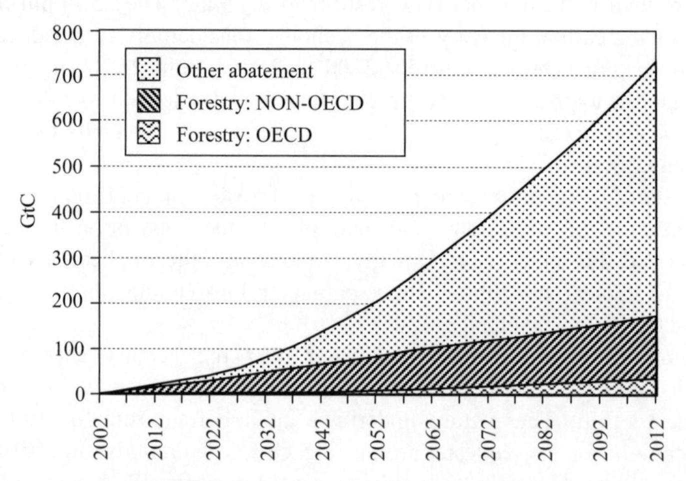

Source: Tavoni et al. (2007).

Figure 5.2 Carbon abatement

the carbon sequestered from 2002 to 2052 of the stabilisation scenario results from reductions in deforestation in just a few regions, namely Latin America, East Asia, and sub-Saharan Africa. Most of this carbon is due to reductions in deforestation. While consideration of policies to reduce deforestation has been shunned in earlier negotiations related to the Kyoto Protocol, they received significant attention as a result of discussions at COP 11 in Montreal.

Focusing on Latin America, East Asia, and Sub-Saharan Africa, where the bulk of deforestation is occurring, around 10.7 million hectares of forestland are estimated to be lost each year (Table 5.2). The carbon incentives in the

Table 5.1 Regional forest carbon sequestration, 2025, 2055, 2095

	2022 MtC/yr	2052	2092
OECD			
USA	42	144	193
OLDEURO	37	82	132
NEWEURO	8	18	29
CAJANZ	31	115	125
Total OECD	118	360	479
NON OECD			
KOSAU	25	27	36
TE	179	117	134
MENA	73	49	31
SSA	270	175	106
SASIA	34	57	32
CHINA	109	155	431
EASIA	451	481	371
LACA	391	326	330
Total Non-OECD	1649	1746	1950
Total Global	1766	2105	2429
C Price	USD57	USD113	USD271

Notes: CAJANZ: Canada, Japan, and New Zealand. KOSAU: Korea, South Africa, and Australia. TE: Transition Economies. MENA: Middle East and North Africa. SSA: Sub-Saharan Africa. SASIA: India and South Asia. EASIA: South East Asia. LACA: Latin America and Caribbean.

Source: Tavoni et al. (2007).

Table 5.2 Net land area change in regions currently undergoing substantial deforestation (Million hectares per year)

	FAO	Projected For		
	(2000–2005)	2002–12	2012–22	2022–32
Latin and Central America	–4.7	–2.3	–0.9	0.2
East Asia	–2.8	–1.2	–0.4	–0.1
Sub-Saharan Africa	–3.2	–2.4	–0.1	0.0
Total	–10.7	–5.9	–1.4	0.1

stabilisation scenario would reduce these losses to around 5.9 million hectares per year during the first decade, and they would essentially halt net forest losses by 2022. While developing policies to reduce deforestation efficiently would undoubtedly be a difficult task, these results suggest that the economic value of making these changes could be substantial.

The overall size of the carbon program increases over the century as carbon prices rise. It increases in both the OECD and the non-OECD regions, but the largest percentage gains occur in the OECD, where the annual carbon sink rises from 118 million t C/yr to 479 million t C/yr. In most non-OECD regions, the strength of the sink is actually declining because there are no longer opportunities to reduce deforestation, and forest growth on large areas of land that were reforested during the century is starting to slow. The one outlier is China, where sequestration expands. Sequestration dynamics in China tend to be more similar to OECD countries because it has large areas of temperate forests that have long growing cycles.

By reducing deforestation and promoting afforestation, a forest carbon sequestration program as part of a stabilisation strategy would have strong impacts on total forestland area in the world, increasing it by 1.1 billion hectares relative to the baseline, or around 0.7 billion hectares above the current area of forests (Table 5.3). The largest share of increased forest area occurs in non-OECD countries. The stabilisation scenario has complex results on timber harvests and prices. Initially, timber is withheld from the market in order to provide relatively rapid forest carbon sequestration through aging timber. As a result, global harvests decline 14.5 percent relative to the BaU in 2022. However, throughout the century, more forests imply a larger supply of timber. By 2092 timber harvests increase by 26 percent. The changes in specific regions heavily depend upon the types of forests (the growth function), the carbon in typical forests (biomass expansion factors), and economic conditions such as prices and costs. In contrast to the area changes, the largest increases in timber harvests (in

Table 5.3 *Change in forestland area and change in annual timber harvests compared to the baseline*

	Million hectares			% Change in ann. harvest		
	2022	2052	2092	2022	2052	2092
OECD						
USA	1.5	23.1	94.2	1.2%	–9.0%	48.5%
OLDEURO	11.5	34.9	51.9	–5.3%	12.1%	0.3%
NEWEURO	2.6	7.8	11.6	–5.3%	12.1%	0.3%
CAJANZ	–4.0	24.5	99.0	–3.8%	–3.3%	167.3%
Total OECD	11.6	90.3	256.7	–3.3%	3.0%	54.1%
NON OECD						
KOSAU	5.1	17.7	49.1	11.3%	34.5%	42.1%
TE	19.0	52.2	102.7	–20.8%	8.9%	–26.1%
MENA	10.3	24.9	38.4	–63.9%	–45.9%	–6.7%
SSA	37.2	90.7	137.0	–70.1%	–52.9%	–9.0%
SASIA	5.2	18.8	32.3	–3.7%	–3.9%	13.0%
CHINA	8.6	41.9	115.4	–20.1%	0.0%	–98.8%
EASIA	25.6	66.0	111.9	–63.3%	–57.2%	–48.9%
LACA	42.9	129.3	262.4	–24.8%	–7.1%	15.5%
Total Non-OECD	153.8	441.5	849.2	–31.9%	–15.4%	–14.9%
Total	165.4	531.8	1105.9	–14.5%	–3.3%	25.9%

Notes: CAJANZ: Canada, Japan, and New Zealand. KOSAU: Korea, South Africa, and Australia. TE: Transition Economies. MENA: Middle East and North Africa. SSA: Sub-Saharan Africa. SASIA: India and South Asia. EASIA: South East Asia. LACA: Latin America and Caribbean.

relative and total terms) occur in OECD countries. OECD countries tend to have many species amenable to producing wood products.

Including forestry management as an abatement strategy has significant general equilibrium effects. As a comprehensive measure of the influence of biological sequestration on the carbon-market, we first examine what happens to the price of carbon when forestry is included into the policy. Figure 5.3 shows the carbon price for the 550 ppm policy throughout the century as found in the original version of the WITCH model (iter1), and after it has been coupled with the forestry model (iter11). Forest sinks substantially lower the cost of CO_2, for example by 40 percent in 2050, making a 550 ppm policy costs as much as a 600pmmv without including forestry. That is, carbon sinks achieve an additional 50 ppm – or equivalently ¼ °C – in 2100 at no extra cost.

To corroborate the idea that forestry can alleviate the compliance to the 550 ppm target, in Figure 5.4 we show the policy costs with and without forestry. Forest sinks are shown to decrease policy costs: in particular, the policy burden is reduced and brought forward to the first half of the century, up to 2050, when the main mitigation action is through avoided deforestation.

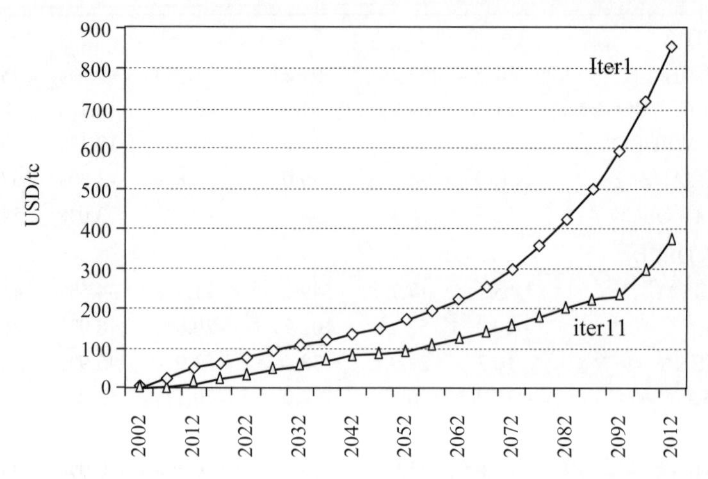

Source: Tavoni et al. (2007).

Figure 5.3 Price of Carbon with (iter11) and without (iter1) forestry

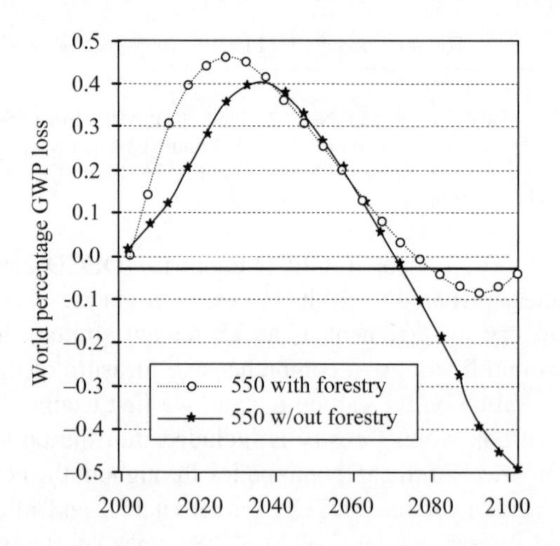

Source: Tavoni et al. (2007).

Figure 5.4 Policy costs with and without forestry

After 2070 the policy-induced benefits from avoided climate damages outweigh the costs of reducing emissions, and this effect is reinforced when forestry is an available mitigation option. All in all, the world policy cost in net present value decreases from 0.2 percent without forestry to 0.1 percent with forestry. This corresponds to a net present value saving to 2100 of almost USD3 USD trillion, which is nearly three times the present value cost of adding the forestry program, estimated at USD1.1 USD trillion.

One might wonder about the distributional effects of including forestry. Two competing effects are at stake: on the one hand forestry will benefit developing countries that are rich in tropical forests, given the role of avoided deforestation. On the other hand, the lower price of carbon will benefit countries that buy carbon-market permits, and disadvantage sellers. Ultimately, the distributional effects will depend on the emissions allocation scheme adopted in the policy. For example, if one assumes that emissions are allocated based on an equal per capita rule, as we do in this chapter, most of the emissions reductions are borne by the developed countries. Lower carbon prices with forestry included in the stabilisation policy improve welfare in OECD countries by reducing their costs (from an undiscounted loss of 0.6 percent without forestry to 0.2 percent with forestry). On the contrary, non-OECD countries tend to be carbon permit sellers, and they have lower revenues when forestry is included as an option, although the difference in revenues is fairly small (from an undiscounted gain of 0.38 percent without forestry to 0.27 percent with forestry). It is worth noting that a different allowance allocation scheme would change the distributional results, but it would not have any impact on the carbon prices as they are determined by the world marginal abatement costs.

One of the key political issues that have hampered the inclusion of forestry in the Kyoto protocol mechanisms is the danger that the emission constraints on the energy system might be relaxed too much: the deployment of clean technologies that can reduce emissions permanently might be delayed, along with the investments in innovation that are needed to make new technologies competitive. This is a justified reason of concern, given the low turnover of energy capital stock, as well as the lengthy process before the commercialisation of advanced technologies. The energy sector description and the endogenous technological change feature of the WITCH model allows to explore the implications that including forestry in the carbon-market has on abatement in the energy sector.

While the climate target dynamically induces larger reductions in energy intensity with respect to the BaU scenario, this reduction is more moderate when we include the forestry abatement option. Indeed, energy intensity remains close to the BaU in the first two to three decades of this century, when avoided deforestation is significantly contributing to abatement, and then approaches the no-forestry path, as the emissions cuts in the energy sector become more predominant. As an example, when REDD is active, the deployment of coal power plants with carbon capture and storage is postponed from 2015 (without forestry) to 2030 (with forestry). Similarly, the share of nuclear power is lower with forestry. Such a setback of low-carbon technologies can be either seen as harmful for the global warming cause, or, more optimistically, as a bridge solution in the interim to develop more consolidated – yet currently uneconomical – technologies.

As WITCH features endogenous technological change via both Learning-by-Doing (LbD) and energy R&D, the effect of REDD on innovation can also be investigated. In Figure 5.5 we show the implications for LbD, including REDD: we plot the percentage variations in the investment costs of wind-and-solar power plants with respect to the BaU case, with and without forestry. Forest sinks hamper the capacity of the 550 ppm policy to induce technological change, as shown by the lower decrease in renewables' costs, due to lower capacity deployment. Moreover, energy R&D investments are decreased by the inclusion of forestry, by roughly 10 percent. Inevitably, the inclusion of forestry to meet a given mitigation target crowds out other abatement options. Although these variations are not large in absolute terms, it would still be efficient to complement the system with policies to foster technological innovation, as this could play a crucial role in hedging against possible future revisions of the climate targets.

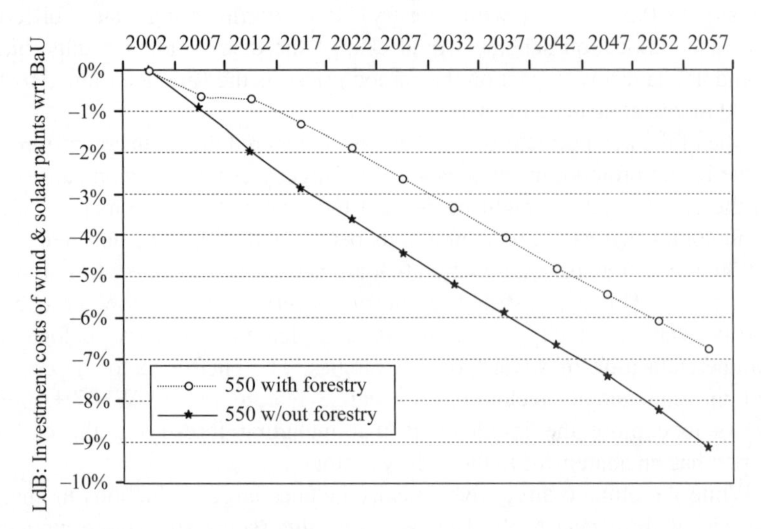

Source: Tavoni et al. (2007).

Figure 5.5 Induced technological change with and without forestry

Hard-link Analysis

The analysis in the previous section relied on immediate and economically efficient participation of all regions in the mitigation effort. Tavoni et al. (2007) modelled a carbon-market system where emission permits are traded giving a more realistic set of targets and geometry of regional participation. The model assumes that short-term climate targets follow recent announcements by different countries over the near to medium term. During the second part of the century, countries' efforts are assumed to start converging to stabilise greenhouse gas concentrations at 550 ppm of CO_2 equivalent. This policy scenario incorporates potential restrictions on trading between industrialised and developing nations during the early years of the market. We also test the effect of allowing the banking of carbon-market permits throughout the century as a tool to increase the efficiency of the policy. Figure 5.6 shows the time profile of carbon emissions for a BaU and a 500 ppm equivalent policy for both OECD and non-OECD groups of countries.

The main results on the role of REDD in the abatement portfolio and the resulting effects on deforestation and climate policy costs are summarised in Table 5.4. REDD represents a relatively important, although declining, source of overall global abatement, particularly with banking, which enables the world to take greater advantage of higher availability of REDD and other cost-effective opportunities in the early periods. The global estimates differ in terms of costs, quantities, and regional distribution of potential, but yield similar aggregate patterns of REDD. In the case of banking, based on the estimates from the GTM and IIASA models respectively, REDD represents between a 19 percent and 20 percent of cumulative abatement by 2020, falling to about 9 percent by 2050 and 4 percent by the end of the century for both models. In contrast, without banking, the contribution of REDD is less than half as much during the first decades and slightly lower throughout the century, representing 3.2 percent and 7.9 percent of cumulative global abatement by 2020, 7.5 percent and 7.2 percent by 2050, and 3.5 percent and 3.6 percent by 2100, for the GTM and IIASA models respectively.

When the WHRC Brazil estimates are examined, reducing deforestation emissions is still a significant source of abatement and actually represents a larger share of total abatement when banking is not allowed as opposed to when banking is possible. This is because REDD in Brazil is estimated to be a relatively cost-effective source of abatement in the early years. This cheap abatement option is largely pursued even without banking, given its limited scale. In the Brazil-only case, REDD represents 5.6 percent (9.4 percent) 2.9 percent (3.1 percent) and 1.6 percent (1.7 percent) share of cumulative abatement by 2020, 2050, and 2100 with (and without banking). As noted earlier, modelling the 'leakage' or displacement of deforestation emissions to

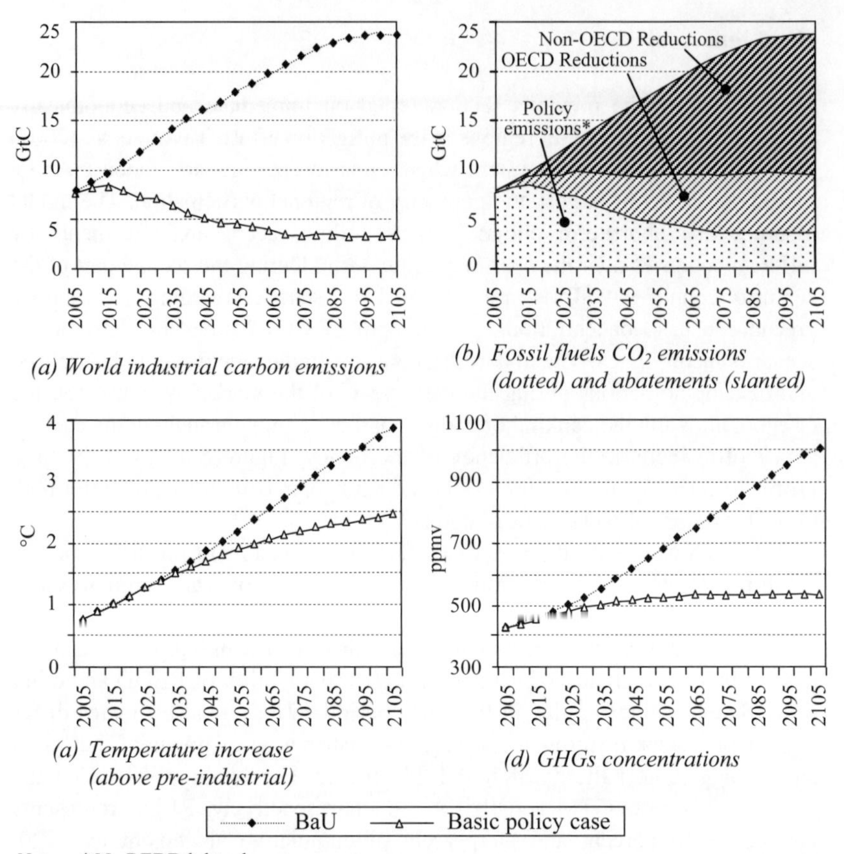

Note: * No REDD ltd trade.

Source: Based on scenarios prepared for Bosetti et al. (2011).

Figure 5.6 Emissions and climate impacts under Business-as-Usual and climate policy scenarios in the hard-link analysis

other parts of the world would reduce the estimated contribution of a Brazil-only REDD scenario.

The flexibility to better optimise the timing of abatement through banking shifts forward the time profile of abatement, in particular when REDD is available. For example, in the IIASA case without banking, 22.6 percent of abatement this century is achieved by 2050, both with and without REDD as a mitigation option. In contrast, when banking is permitted, the share of the abatement achieved by 2050 rises to 24.1 percent and 23.9 percent in the case with and without REDD, respectively. The time profile of emission reductions has important implications when considering tipping points and abrupt changes in the climate state (as the

Table 5.4 *Estimated cumulative impact of REDD on abatement share, deforestation emissions, and global policy costs, by time period*

Variable (REDD scenario below)		2010–19	2010–49	2010–99
Cumulative Share of REDD in Global Abatement				
With Banking	WHRC Brazil	5.6%	2.9%	1.6%
	Global Timber Model	19.3%	9.2%	4.1%
	IIASA Model	19.8%	8.7%	4.1%
Without Banking	WHRC Brazil	9.4%	3.1%	1.6%
	Global Timber Model	3.2%	7.5%	3.5%
	IIASA Model	7.9%	7.2%	3.6%
Cumulative Reductions in Emissions from Deforestation				
With Banking	WHRC Brazil	–15.7%	–21.7%	–29.6%
	Global Timber Model	–72.2%	–87.8%	–90.9%
	IIASA Model	–50.3%	–63.9%	–68.4%
Without Banking	WHRC Brazil	–11.3%	–20.4%	–28.9%
	Global Timber Model	–4.9%	–64.3%	–78.1%
	IIASA Model	–8.2%	–47.6%	–59.9%
Cumulative Reductions in Loss of Gross World Product (GWP) [a]				
With Banking	WHRC Brazil	–7.2%	–8.5%	–9.9%
	Global Timber Model	–7.6%	–17.7%	–21.4%
	IIASA Model	–10.6%	–19.8%	–22.9%
Without Banking	WHRC Brazil	–6.9%	–7.8%	–11.1%
	Global Timber Model	–13.3%	–18.7%	–24.0%
	IIASA Model	–15.4%	–17.2%	–22.2%

Note (a) Estimates are based on a 5 percent discount rate and are relative to the no-REDD policy case.

shutting down of the thermohaline circulation), hence options that anticipate emission reductions could also bear the benefit of reducing these risks.

In the base case, without REDD and without banking, the permit price is estimated to begin around USD4/t CO_2 equivalent, rising sharply to USD36 by 2020 and to almost USD400 by 2050. When banking is allowed, the price trajectory is flattened, with higher prices in the early years and lower prices in the later years. With banking, the price rises to USD87 over 2015–19 but only to USD330 over 2045–49.[4] The availability of REDD mitigation moderates the level of prices. In the no-banking case, REDD has negligible impacts on the price prior to 2020, as other abatement opportunities are relatively plentiful and the quantities of international trading are restricted. The estimated effect of linking REDD to a global carbon-market is to significantly and rapidly reduce global deforestation emissions (see the middle set of rows in Table 5.2). In the banking cases, tropical deforestation emissions decline by an estimated 16 percent (WHRC Brazil), 72 percent (GTM), and 50 percent (IIASA) by 2019 and by 22 percent (WHRC Brazil), 88 percent (GTM), and 64 percent (IIASA) by 2049. When global REDD is modelled, global deforestation emissions decrease such that the global forest sector becomes a net sink (negative net emissions) by the middle of the century.

The basic policy without REDD results in cumulative GWP losses over 2010–2099 of 2.5 percent (1.8 percent) at a 3 percent (5 percent) discount rate.[5] The added flexibility in the timing of abatement due to banking lowers these losses to 2.1 percent (1.6 percent) at a 3 percent (5 percent) discount rate. Despite the restrictions on overall trading and the modest relative contribution to total global abatement, REDD decreases the estimated costs of meeting a global climate target, with the impact depending on the estimated potential and the availability of banking. As shown in Table 5.2, in the case without banking, REDD lowers overall policy costs for the century by 11 percent, 24 percent, and 22 percent based on the WHRC-Brazil, GTM, and IIASA estimates, respectively. Irrespective of whether REDD is available as a mitigation option or not, policy costs are lower when banking is allowed. With banking, REDD reduces the costs by 10 percent, 21 percent and 23 percent in the WHRC-Brazil, GTM, and IIASA cases, respectively.[6] These estimated effects of a global REDD scenario have a greater impact on policy costs than the availability and or absence of carbon-free and low-carbon technologies, as documented in various publications (Bosetti et al., 2009).

The cost-savings from introducing REDD indicate that a more stringent target is feasible at the same costs of a mitigation policy without REDD. Focusing on the IIASA case, we simulate a series of more stringent scenarios in which we proportionally escalate all regional efforts during the second part of the century. In this way, we identify a scenario with equivalent costs to the basic policy scenario without REDD, that is, implying a 2.5 percent estimated decline in Global World Product (GWP). This exercise suggests that allowing the link with REDD makes it possible to achieve a scenario resulting in a reduction of emissions of 20 ppm of CO_2-equivalent below the base policy case for the same amount of discounted costs.

As indicated in the previous section, the inclusion of REDD as an abatement option may impact technology development in the energy sector. Our results indicate a generally negative but modest overall effect of REDD on energy technology R&D and low-carbon technology investments, although we also find small positive impacts for some particular technology categories. The effect of REDD on technology investments and resulting innovation follows two channels. REDD makes it possible to attain the stabilisation target while slightly relaxing the pressure to reduce energy emissions. In particular, REDD allows for a 2 percent, 8 percent and 10 percent increase in the cumulative emissions from the energy sector over 2010–2049, in the WHRC-Brazil, GTM, and IIASA REDD scenarios respectively. Figure 5.7 shows how REDD slightly reduces the need for improvements in carbon efficiency both in 2030 and 2050, which expands the market for fossil fuel technologies, compared to the case without mitigation from REDD. Under the BaU, energy efficiency is estimated to slightly improve, while carbon efficiency is estimated to progressively decline relative to the past 30 years, primarily due to the increase in the use of coal in developing countries. Irrespective of REDD, both energy and carbon efficiency improve dramatically under the climate policy cases.

By granting some leeway to the fossil fuel sector, linking REDD to the carbon-market decreases investments in the development of renewable (wind and solar) and nuclear energy sources, as well as in energy-intensity R&D.

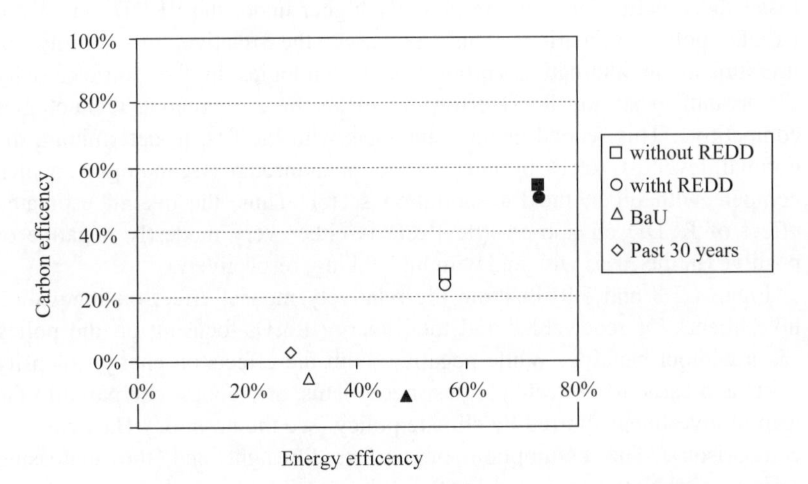

Note: Projections for the year 2030 and 2050 are shown in white and black, respectively. Improvements in energy and carbon intensities are with respect to the base year under the baseline and in the two policy cases with and without REDD (IIASA scenario, without banking). Energy and carbon efficiency improvement in the last 30 years are also reported for comparison.

Source: Based on scenarios prepared for Bosetti et al. (2011).

Figure 5.7 Relative improvements in energy and carbon intensity

The estimated investment reductions are of the order of 3–10 percent in the global scenarios compared to the non-REDD climate policy case. The order of magnitude is modest relative to the substantial estimated increases in all these investments as a result of the climate policy, regardless of REDD, compared to the no-policy case (ranging from 60 to 80 percent for solar/wind and energy-intensity R&D).

While there is a trade-off between relaxing the constraint on energy sectors emissions and the size of investments in renewable, nuclear, and energy-intensity R&D, a different story holds true for investments in integrated gasification combined cycle (IGCC) plus carbon capture and storage (CCS) technologies. This technology is not entirely carbon-free given an estimated emissions capture rate of 90 percent. For this reason, relaxing the cap on energy emissions improves the economic competitiveness of the IGCC plus CCS option. Over the medium run, investments in IGCC plus CCS technologies thus expand slightly from the introduction of REDD (in the order of 1 percent in the global scenarios).

The second channel through which REDD affects the estimated patterns of technology investments is through the impact on fossil fuel prices. As REDD allows greater flexibility in reducing fossil fuel consumption, the prices of fossil fuels, particularly oil, are slightly higher under the REDD versus no-REDD policy scenarios. This increases the relative profitability of investments in alternative carbon-free technologies in the non-electricity sector and boosts the R&D efforts to make these alternative technologies competitive. This second channel interacts with the first in determining the optimal level of R&D in non-electric breakthrough technologies, which compete with oil in the transportation sector. Thus, the overall estimated effect of REDD on non-electric R&D is either very modestly negative or positive (in the cases with and without banking, respectively).

Figures 5.8 and 5.9 illustrate the relatively modest effect on innovation investments for renewables and total energy R&D, focusing on the policy cases without banking. While negative, both the effects on energy intensity R&D and carbon-free technologies investments, are modest compared to the leap in investment spurred by climate policy (see the no-policy BaU case for comparison). The assumption of perfect foresight and the increasing emissions reduction required by the policy architecture call for significant investments in new technologies and REDD only moderately decreases the required effort.

When the policy targets are anticipated, the effect of REDD on innovation is dramatically reduced under more stringent climate targets. In particular, focusing on the IIASA no-banking case, if we consider a scenario where global emissions reductions are increased after 2050 to achieve 20 ppm less by 2100, REDD implies a reduction in nuclear investments over the first half

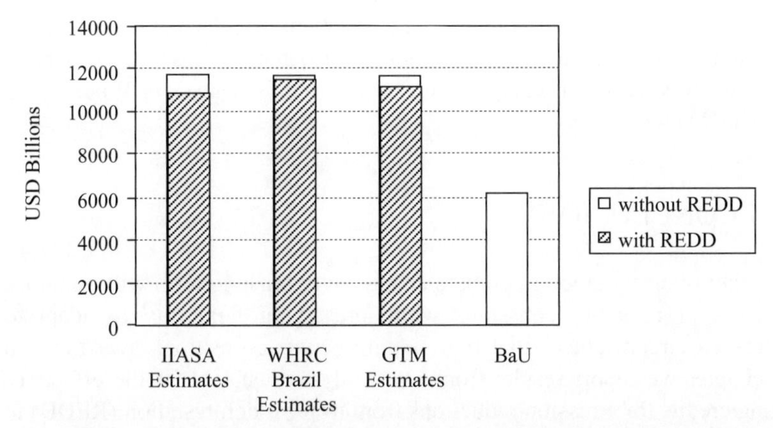

Note: The entire height of each column indicates the case without REDD, while the shaded portion indicates the case with REDD, according to the estimates from the different models under the no-banking cases. Business-as-Usual projections without any climate policy are provided for comparison.

Source: Based on scenarios prepared for Bosetti et al. (2011).

Figure 5.8 Cumulative investments in carbon-free technologies (2010–2050: Wind plus Solar and Nuclear) under no-banking cases

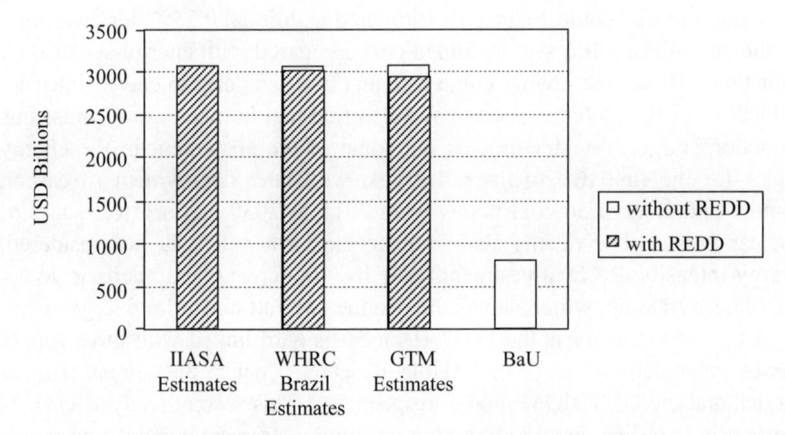

Note: The entire height of the column indicates the case without REDD, while the shaded portion indicates the case with REDD, according to the estimates from the different models under the no-banking cases. An exception is the IIASA estimates where the cumulative investment with REDD is just slightly higher than in the case without REDD. Business-as-Usual (BaU) projections without any climate policy are provided for comparison.

Source: Bosetti et al. (2011).

Figure 5.9 Cumulative investments in total energy R&D (2010–2050) under no-banking cases

of the century of 1 percent, rather than 6.6 percent. Similarly, the decrease in wind and solar investments falls from about 7 percent to less than 4 percent while the contraction of energy intensity R&D falls from over 9 percent to less than 7 percent.

5.6 CONCLUSIONS

Efficient policies to address climate change over the long-term will minimise the costs of reducing emissions while preserving flexibility to adapt to unforeseen circumstances that may require course corrections over time. In this chapter we report results from two analyses that look at the effects of linking credits for emission reductions from tropical deforestation (REDD) to a global carbon-market based on a dynamic integrated assessment framework, which explicitly models induced technological change in the energy sector.

The first analysis, where the WITCH model is soft-linked to the Global Timber Model (GTM) of the forestry sector, indicates that forestry mitigation enables an atmospheric target of 550 CO_2 parts per million by volume (ppm) for the same total cost as a 600 ppm target without forestry mitigation. The estimated net cost savings of two USD trillion (40 percent of policy costs in discounted terms) could finance an estimated additional 0.25°C less warming by the end of the century at no added cost compared with energy-sector only reductions. These cost savings come with an estimated delay in energy-intensity reductions of the global economy. We also find that in meeting the emissions reductions target, forestry crowds out some of the abatement in the energy sector for the first two to three decades, with later deployment of carbon capture and storage on coal plants from 2015 to 2030 and a lower share of nuclear power. The results also indicate that, when REDD is considered, energy-intensity R&D investments fall by 10 percent and there is lower Learning-by-Doing, which delays the decline of wind-and-solar energy costs.

In the second analysis the WITCH model is hard-linked with three supply curves generated by the Woods Hole Research Center, the Global Timber Model, and the GLOBIOM model, respectively. This second analysis extends the first in two directions. First it examines more stringent stabilisation policy (500 ppme). In doing so, it incorporates expected patterns of global participation as well as institutional features considered likely, such as limits on initial international trading and potential for permit banking. It also uses scenarios to explore the effect of REDD on policy flexibility to tighten targets in the future. Our research confirms that integrating REDD into global carbon-markets can provide significant incentives for reducing deforestation, while lowering the costs of global climate change protection. We find that the cost savings from

REDD have only modest trade-offs in terms of reduced clean energy innovation. We find that including REDD as a mitigation option generally reduces investments in lower-carbon cleaner energy technologies over the next four decades by a maximum of 10 percent in the case of for energy-intensity R&D investments over 2010–49. This crowding out effect is not uniform across technologies. REDD could provide a small estimated boost on the development of some efficient but fossil-based technologies, including IGCC-CCS. Moreover, while reduced clean energy innovation could, in principle, hinder future efforts to reduce emissions, our estimates suggest such effects would not compromise the flexibility of future climate policies under REDD. On the contrary, we find synergies between REDD and the possibility of banking that, overall, increase policy resilience by reducing the costs of a potential future increase in the stringency of climate policy. Introducing REDD lowers the total costs of the stabilisation policy over this century by an estimated 10–23 percent depending on different model estimates for reductions from Brazil only as well as all tropical countries. By the same token, we explored an increasingly stringent scenario to the point where costs with REDD matched those of the basic policy scenarios without REDD. We find that REDD could enable additional reductions of approximately 20 ppm of CO_2-equivalent by the end of the century with no added discounted costs compared to the basic policy case.

Results from this second study indicate that global REDD contributes about 9 percent to 7 percent of total cumulative global abatement throughout the first half of the century, with and without banking. This is lower than what was estimated with the first study, where non-OECD and global forests could, respectively, contribute one-quarter and one-third of cost-effective abatement by mid-century, with forestry overall defraying the costs of 50 ppm additional reductions by 2100. The large potential for forestry found in the first analysis builds partly on the moderate target that is considered (550 ppm CO_2-only compared to roughly 450 ppm CO_2-only in the second study) and partly on the assumption of full availability of mitigation opportunities (from avoided deforestation plus afforestation, reforestation, and changes in forest plantation management worldwide).

According to the second study, REDD generally reduces the total portfolio of investments and R&D of low-carbon energy technologies by about 1–10 percent, depending on the REDD estimates and policy case. These effects are relatively modest compared to the overall impacts of the policy versus no-policy cases, regardless of REDD. The energy-intensity R&D impacts in the first study correspond to the upper end of this broader range.

To conclude, linking REDD to the carbon-market, even gradually (starting from Brazil, where the level of preparedness is very high, and then including other countries), could be extremely beneficial. It could reduce the costs of mitigation action and provide incentives to developing countries to be part of an international agreement (see also Chapter 3 on international agreements in this book). Co-benefits from avoided deforestation in terms of biodiversity preservation, ecosystem services, and livelihoods for communities would only add economic rationale to the proposed approach.

NOTES

1. This chapter is based on the following papers: Bosetti et al. (2011) and Tavoni et al. (2007).
2. The integrated assessment model and the forestry model both run independently, but each takes as input information from the other model and the iterations stop when the behaviour of each model converges.
3. Reduce forms forestry supply curves are included within the WITCH framework.
4. Estimates of stabilisation costs computed applying WITCH, are higher than those reported by IPCC (2007). Marginal abatement costs crucially depend on assumptions about availability and penetration of carbon-free technologies in the electric and non-electric sectors. This is particularly true for more stringent scenarios where almost complete decarbonisation of the economy is required by the end of the century. In WITCH, multiple carbon-free alternatives are modelled for the electricity sector, whereas new technologies become competitive in the non-electricity sectors only through large investments in R&D. In addition, the diffusion processes for new technologies are modelled to mimic the time required in order to undertake the extensive infrastructural changes. Finally, WITCH features a non-cooperative representation of knowledge creation and diffusion processes. All these factors are at the basis of these higher estimates.
5. These results are robust to the level of forestry emissions incorporated in the baseline.
6. The gains in policy costs measured in discounted terms depend upon the choice of the discount rate, even though only marginally, as a large part of the cost reduction accrues during the first part of the century. Here we are using a discount rate of 5 percent, but if we were to choose a 3 percent discount rate, the cost savings would be increased to 11 percent, 21 percent, and 25 percent in the three cases with banking.

REFERENCES

Anger, N. and J. Sathaye (2008), 'Reducing deforestation and trading emissions: economic implications for the post-Kyoto market', Discussion Paper No. 08-016, Mannheim, Germany: Center for European Economic Research.

Bosetti, V., C. Carraro, R. Duval, A. Sgobbi and M. Tavoni (2009), 'The role of R&D and technology diffusion in climate change mitigation: New perspectives using the WITCH Model', OECD Economics Department Working Paper No. 664, OECD publishing, doi:10.1787/227114657270.

Bosetti, V., R. Lubowski, A. Golub and A. Markandya (2011), 'Linking reduced deforestation and a global carbon market: Implications for clean energy technology and policy flexibility', *Environment and Development Economics*, **16**(04), 479–505, doi: 10.1017/S1355770X10000549.

Calvin, K., P. Patel, A. Fawcett, L. Clarke, K. Fisher-Vanden, J. Edmonds, S.H. Kim, R. Sands and M. Wise (2009), 'The distribution and magnitude of emissions mitigation costs in climate stabilization under less than perfect international cooperation: SGM results', *Energy Economics*, **31**, S187–S197.

Chomitz, K. (2006), 'Policies for national-level avoided deforestation programs: A proposal for discussion', background paper for policy research paper on tropical deforestation.

Dinan, T.M. and P.R. Orszag (2008), 'It's about timing', *The Environmental Forum*, **25**(6), 36–39.

Dixon, A., N. Anger, R. Holden and E. Livengood (2009), 'Integration of REDD into the international carbon market: Implications for future commitments and market regulation', Report prepared for The New Zealand Ministry of Agriculture and Forestry by M-co Consulting, New Zealand, and Centre for European Economic Research (ZEW), Germany

EC (Commission of the European Communities) (2008), 'Addressing the challenges of deforestation and forest degradation to tackle climate change and biodiversity loss', Communication from the commission to the European Parliament, the Council, the European Economic and Social Committee and the Committee of the Regions, COM(2008) 645/3, Brussels, Belgium.

Edmonds, J.A., J. Clarke, J. Dooley, S.H. Kim, R. Izaurralde, N. Rosenberg and G. Stokes (2003), 'The potential role of biotechnology in addressing the long-term problem of climate change in the context of global energy and ecosystems', in J. Gale and Y. Kaya (eds), *Greenhouse Gas Control Technologies*, Amsterdam: Pergamon, pp. 1427–1432.

Eliasch, J. (2008), 'Climate change: Financing global forests', Office of Climate Change, UK.

Fisher, B., N. Nakicenovic, K.Alfsen, J. Corfee Morlot, F. de la Chesnaye, J.-C. Hourcade, K. Jiang, M. Kainuma, E. La Rovere, A. Matysek, A. Rana, Keywan Riahi, R. Richels, S. Rose, D. Van Vuuren, R. Warren (2007), 'Issues related to mitigation in the long-term context', in IPCC (Intergovernmental Panel on Climate Change), *Climate Change 2007: Mitigation, Contribution of Working Group III to the Fourth Assessment Report of the Intergovernmental Panel on Climate Change* [B. Metz, O.R. Davidson, P.R. Bosch, R. Dave, L.A. Meyer (eds)], Cambridge, UK and New York, NY, USA: Cambridge University Press, pp. 169–250.

Golub, A., S. Fuss, J. Szolgayova and M. Obersteiner (2008), 'Effects of low-cost offsets on energy investment: New perspectives on REDD', Nota di Lavoro 17.2009, Milan: Fondazione Eni Enrico Mattei.

Golub, A., N. Greenberg, J.A. Anda and J.S. Wang (2009), 'Low-cost offsets and incentives for new technologies', in A. Golub and A. Markandya (eds), *Modeling Environment-Improving Technological Innovations under Uncertainty*, New York, NY: Routledge, pp. 309–327.

Government of Brazil (2008), 'National plan on climate change: Executive summary', Interministerial Committee on Climate Change, Decree No. 6263 of November 21, 2007, Brasilia, Brazil.

Gusti, M., P. Havlik and M. Obersteiner (2008), 'Technical description of the IIASA model cluster', International Institute for Applied Systems Analysis (IIASA).

Hansen, M., S. Stehman, P. Potapov, T. Lovaland, T. Townshend, R. DeFries, K. Pittman, B. Arunarwati, F. Stolle, M. Steininger, M. Carroll and C. Di Miceli (2008), 'Humid tropical forest clearing from 2000 to 2005 quantified by using multitemporal and multiresolution remotely sensed data', *Proceedings of the National Academy of Sciences*, **105**(27), 9439–9444.

Hertel, T., S. Rose and R.S.J. Tol (2009), 'Land use in computable general equilibrium models: An overview', in T. Hertel, S. Rose and R.S.J. Tol (eds), *Economic Analysis of Land Use in Global Climate Change Policy*, New York, NY: Routledge, pp. 3–30.

IPCC (Intergovernmental Panel on Climate Change) (2007), *Climate Change 2007: Impacts, Adaptation and Vulnerability. Contribution of Working Group II to the*

Fourth Assessment Report of the Intergovernmental Panel on Climate Change [Parry, M.L., O.F. Canziani, J.P. Palutikof, P.J. van der Linden and C.E. Hanson, (eds)], Cambridge, UK and New York, NY, USA: Cambridge University Press.

Kindermann, G., M. Obersteiner, B. Sohngen, J. Sathaye, K. Andrasko, E. Rametsteiner, B. Schlamadinger, S. Wunder and R. Beach (2008), 'Global cost estimates of reducing carbon emissions through avoided deforestation', *Proceedings of the National Academy of Sciences*, **105**(30), 10302–10307.

Murray, B.C., R. Lubowski and B. Sohngen (2009), 'Including international forest carbon incentives in climate policy: Understanding the economics', Nicholas Institute Report NI R 09-03, Durham, NC: Nicholas Institute for Environmental Policy Solutions, Duke University.

Murray, B.C, R.G. Newell and W.A. Pizer (2009), 'Balancing cost and emissions certainty: An allowance reserve for cap-and-trade', *Review of Environmental Economics and Policy*, **3**(1), pp. 84–103.

Nepstad, D., B.S. Soares-Filho, F. Merry, A. Lima, P. Moutinho, J. Carter, M. Bowman, A. Cattaneo, H. Rodrigues, S. Schwartzman, D.G. McGrath, C.M. Stickler, R. Lubowski, P. Piris-Cabezas, S. Rivero, A. Alencar, O. Almeida and O. Stella (2009), 'The end of deforestation in the Brazilian Amazon', *Science*, **326**, 1350–1351.

Nepstad, D., B.S. Soares-Filho, F. Merry, P. Moutinho, H. Oliveira Rodrigues, M. Bowman, S. Schwartzman, O. Almeida and S. Rivero (2007), 'The costs and benefits of reducing deforestation in the Brazilian Amazon', Woods Hole, MA: The Woods Hole Research Center.

Obersteiner M., Ch. Azar, P. Kauppi, K. Möllersten, J. Moreira, S. Nilsson, P. Read, K. Riahi, B. Schlamadinger, Y. Yamagata, J. Yan and J.-P. van Ypersele (2001), 'Managing climate risk', *Science*, **294**(5543), 786–787.

Pacala, S. and R. Socolow (2004), 'Stabilization wedges: Solving the climate problem for the next 50 years with current technologies', *Science*, **305**, 968–972.

Piris-Cabezas, P. and N. Keohane (2008), 'Reducing emissions from deforestation and forest degradation: Implications for the carbon market', Washington, DC: Environmental Defense Fund.

Ronneberger, K., M. Berrittella, F. Bosello and R.S.J. Tol (2009), 'KLUM@GTAP: Spatially explicit, biophysical land use in a computable general equilibrium model', in T. Hertel, S. Rose and R.S.J. Tol (eds), *Economic Analysis of Land Use in Global Climate Change Policy*, New York, NY: Routledge, pp. 304–338.

Rose, S., H. Ahammad, B. Eickhout, B. Fisher, A. Kurosawa, S. Rao, K. Riahi and D. van Vuuren (2008), 'Land in climate stabilization modeling: Initial observations', EMF Report 21, Energy Modeling Forum, Stanford University.

Sohngen, B. and R. Sedjo (2006), 'Carbon sequestration in global forests under different carbon price regimes', *Energy Journal*, **27**,109–126.

Sohngen, B., R. Mendelsohn and R. Sedjo (1999), 'Forest management, conservation, and global timber markets', *American Journal of Agricultural Economics*, **81**(1), 1–13.

Sohngen, B. and R. Mendelsohn (2003), 'An optimal control model of forest carbon sequestration', *American Journal of Agricultural Economics*, **85**(2), 448–457.

Sohngen, B. and R. Mendelsohn (2006), 'A sensitivity analysis of carbon sequestration', in M. Schlesinger (ed.), *Human-Induced Climate Change: An Interdisciplinary Assessment*, Cambridge: Cambridge University Press, pp. 227–237.

Tavoni, M., B. Sohngen and V. Bosetti (2007), 'Forestry and the carbon-market response to stabilise climate', *Energy Policy*, **35**(11), 5346–5353.

United Nations Framework Convention on Climate Change (UNFCCC) (2009), 'Draft decision -/CP.15. Decision by the President. Copenhagen Accord', FCCC/CP/2009/L.7/ available at: http://unfccc.int/resource/docs/2009/cop15/eng/l07.pdf

US Climate Action Partnership (USCAP) (2009), 'Key findings from the economic analysis for the USCAP blueprint for legislative action', United States Climate Action Partnership, Washington, DC. http://www.pewclimate.org/uscap/economic-modeling

Winjum, J.K., S. Brown and B. Schlamadinger (1998), 'Forest harvests and wood products: Sources and sinks of atmospheric carbon dioxide', *Forest Science*, **44**, 272–284.

Wise, M., K. Calvin, A. Thomson, L. Clarke, B. Bond-Lamberty, R. Sands, S. Smith, A. Janetos and J. Edmonds (2009), 'Implications of limiting CO_2 concentrations for land use and energy', *Science*, **324**(5931), 1183–1186.

6. Adaptation and Mitigation: What is the Optimal Balance?

Enrica De Cian

6.1 INTRODUCTION[1]

The latest rounds of international negotiations in Copenhagen and Cancun have witnessed the emergence of a bottom-up strategy in which countries unilaterally decide the mitigation effort to propose. This approach has led to a divide between proposed commitments and what is required to achieve ambitious global environmental goals, whose integrity might actually be put at risk. Cancun made significant progress by bringing the essential elements of Copenhagen back into the UNFCCC framework. Examples are the emission reduction pledges, the Green Climate fund, the Adaptation Framework, the Technology Mechanism. However the feasibility of achieving the 2°C target remains uncertain. This objective, for the first time explicitly mentioned in a UNFCCC document, could be undermined by weak national emission reduction targets, indeed leaving room for not yet defined negative effects. In this context, adaptation obtains a defined and unique role: closing the gap between what needed to be done and what has been done. Hence, policy for adaptation needs more attention and policy makers should complement strong mitigation policy.

To some extent, people can avoid some of the damages caused by climate change by adopting different behaviours, driven by market signals. For example, farmers can choose plants that thrive in the heat or new houses can be designed to deal with warmer temperatures. This form of autonomous adaptation induced by market price signals is also referred to as market-driven adaptation. Socio-economic systems have a large potential to adapt to climate change, but market action might not be sufficient. First of all, market-driven adaptation works well if markets function correctly. Therefore, there are some forms of damage that cannot be addressed by markets, but require some degree of cooperation and the intervention of policy makers. This can be done either locally or internationally. A classic example of adaptation investments that are a local public good are investments to protect against sea-level rise. Second, climate change impacts have an equity-adverse effect and hit poor countries relatively more severely. Climate change adaptation needs will increase over time, especially in developing countries, which will also require investing in institutional and adaptive capacity. The real challenge for adaptation, therefore, lies in tackling climate change impacts in developing countries. In addition, regional patterns of climate change damages are not related to the geography of historical responsibilities. Hence, there is a need for international cooperation on adaptation as well, though for a different reason than mitigation. In this context, the following key questions need to be addressed: how should resources be allocated between mitigation and adaptation over time? How should the equity-adverse impact of climate change be addressed? What are the key priorities to be addressed by adaptation policies and adaptation funds?

Prior studies have explored these questions in a cost-benefit framework. Adaptation is modelled as an aggregated strategy fostered by some form of planned spending, which can directly reduce climate change damage. The pioneering contribution in this field is Hope et al. (1993), who proposed the first effort to integrate mitigation and adaptation into the PAGE Integrated Assessment Model. PAGE, however, defines adaptation exogenously and therefore it cannot determine the optimal characteristics of a mitigation and adaptation portfolio. The first assessments of the optimal mix of adaptation and mitigation where both mitigation and adaptation are endogenous were proposed by Bosello (2008), Bosello et al. (2010) and de Bruin et al. (2009a, 2009b). All these studies conclude that adaptation and mitigation are strategic complements: the optimal policy consists of a mix of adaptation measures and investments in mitigation, both in the short and long-term, even though mitigation will only decrease damages in later periods. All authors also highlight the trade-off between the two strategies: because resources are scarce, investing more into mitigation implies fewer resources for adaptation. Moreover, successful adaptation reduces the marginal benefit of mitigation and a successful mitigation effort reduces the damage to which it is necessary to adapt. This, again, explains the trade-off between the two strategies. However, the second effect is notably weaker than the first one. Mitigation, especially in the short-medium term, lowers only slightly the environmental damage and therefore does little to decrease the need to adapt. Finally, all the aforementioned studies stress that adaptation is a more effective option to reduce climate change damage, especially if agents have a strong preference for the present (high discount rates), or early climate damages are expected. This outcome depends on the cost and benefit functions driving the decision to spend on mitigation and adaptation, which are based on the standard damage functions used in most integrated assessment models, for instance the one from Nordhaus' DICE/RICE models. These damage functions include at best, extreme, but not catastrophic events, and no uncertainty. As a matter of fact, the recent outcomes of international negotiations suggest that more ambitious mitigation policies might be considered, following a precautionary principle argument. Suppose policy makers succeed at implementing a global mitigation policy to stabilise GHG concentrations at 550 ppme by the end of the century. Would adaptation still be justified in the presence of a strong mitigation effort? What would the optimal mix between adaptation and mitigation be when climate policy aims at stabilising GHG concentrations at 550 CO_2-eq?

This chapter aims at shedding light on the two issues raised in the introduction: what is the optimal balance between adaptation and mitigation under cost-benefit considerations and how does this balance change under a stabilisation target. The results have been extracted from a recent body of research performed with the model version of WITCH upgraded to model adaptation, the AD-WITCH model. The structure of this chapter is as follows. Section 6.2 describes the AD-WITCH model and its calibration. Section 6.3 deals with the optimal balance between adaptation and mitigation in a cost-benefit setting, while Section 6.4 does a similar analysis but in a cost-effective scenario. Section 6.5 presents some regional considerations and Section 6.6 concludes.

6.2 THE AD-WITCH MODEL

Modelling adaptation strategies in numerical models for the analysis of mitigation policies makes it possible to quantify adaptation needs under different mitigation and climate change damage scenarios. However, it is also a challenging task. A major difficulty relates to the different nature of mitigation and adaptation. The former are typically studied with international macroeconomic models. On the contrary, adaptation activities and policies often take the form of project-based activities with a local, site-specific relevance and have been addressed mostly with a microeconomic perspective. Reconciling the two views is problematic, but at the same time necessary. To derive strategic long-term policy insights, the interaction between adaptation and mitigation must be analysed from a macroeconomic angle. The AD-WITCH extends the WITCH model to link adaptation, mitigation, and climate change damage within an integrated assessment framework of the world economy.

In the model, adaptation consists of a portfolio of macro-strategies that describe specific features of adaptation measures. Anticipatory adaptation, reactive adaptation, and investment in adaptive capacity take the form of dedicated investments or expenditure flows. When implemented, they decrease climate change damages, but at a cost. Adaptation competes with mitigation and other investments in the process of utility maximisation. Unlike WITCH, the AD-WITCH model separates residual damage from adaptation expenditures, which become policy variables and are optimally set with all other choice variables in the model, for example, investments in physical capital, R&D, and energy technologies. The next section describes in detail the adaptation module while the following one explains the calibration procedure.

The Adaptation Module

The AD-WITCH model links adaptation, mitigation, and climate change damage within the WITCH model. In order to make adaptation comparable to mitigation, the large number of possible adaptive responses has been aggregated into two broad expenditure categories: adaptive capacity building and adaptation activities. Expenditure in adaptive capacity building is further divided into a generic and a specific component. Expenditure in adaptation activities includes anticipatory and reactive adaptation.

A well-developed adaptive capacity is key to the success of adaptation strategies. AD-WITCH includes this component through two variables: generic and specific adaptive capacity building. Generic adaptive capacity building is linked to the overall level of economic and social development of a region. The degree of economic development affects the final impact of climate change on the economic system: for example, a high-population-growth and low-income-per-capita region is more prone to suffer from climate change than a low-population, high-income-per-capita region (IPCC, 2007; Parry, 2009). Specific adaptive capacity building refers to all dedicated investments that are specifically targeted at facilitating adaptation activities. Examples within this category are: the improvement of meteorological services and of early warning systems, the development of climate modelling and impact assessment, and, above all, technological innovation for adaptation purposes.

Anticipatory adaptation gathers all the measures where a stock of defensive capital must already be operational when the damage materialises. A typical example of these activities is coastal protection. Anticipatory adaptation is characterised by some economic inertia as investments in defensive capital take some time before translating into effective protection capital. Therefore, investments must begin before the damage occurs, and, if well designed, become effective in the medium, long-term.

By contrast, reactive adaptation describes the actions that are put in place when climate related damages effectively materialise. Examples of reactive actions are expenditures for air conditioning or treatments for climate-related diseases. These actions must be undertaken period-by-period to accommodate damages not avoided by anticipatory adaptation. They need to be constantly adjusted to changes in climatic conditions. An adaptation tree assembles these adaptation strategies into a sequence of nested CES functions, described in detail in Bosello et al. (2010, 2013) (see also www.witchmodel.org).

Calibration of the AD-WITCH Model

The first major contribution to the literature is the integration of different adaptation strategies in a unified framework. In contrast, previous studies focussed either on reactive (de Bruin et al., 2009a, 2009b) or anticipatory measures (Bosello, 2008), and neglected the role of adaptive capacity building (Bosello et al., 2010). A second novel feature of the model is an updated calibration of macro-regional adaptation costs and effectiveness. Table 6.1 summarises adaptation costs, adaptation effectiveness, and total climate change damages, together with the calibrated values, at the calibration point, when CO_2 concentration doubles.

In the calibration procedure, Bosello et al. (2010) integrate the original database of the WITCH model with Nordhaus and Boyer (2000) and Agrawala and Fankhauser (2008), which provide the most recent and complete assessment on costs and benefits of adaptation strategies. Three major points deserve to be mentioned. First, we gather new information on climate change damages consistent with the existence of adaptation costs. We then calibrate AD-WITCH on these new values and not on the original values of the WITCH model. Second, due to the optimising behaviour of the AD-WITCH model, when a region gains from climate change, it is impossible to replicate any adaptive behaviour and positive adaptation costs in that region. Accordingly, when WITCH data show gains from climate change, we refer to Nordhaus and Boyer (2000) results. If both sources report gains (as in the case of Transition Economies, TE) we impose a damage level originating an adaptation cost consistent with the observations. Third, the calibrated total climate change costs are reasonably similar to the reference values. The main explanation is that consistency needs to be guaranteed across three interconnected items: adaptation costs, total damage, and protection levels. Adaptation costs and damages move together. For instance, it is not possible to lower adaptation costs in Western Europe (WEURO) to bring them closer to their reference value without decreasing total damage, which is already lower than the reference. Although we are fully aware of these shortcomings, we also recognise that the quantitative assessment of adaptation costs and benefits is still at a pioneering stage and that some sectors and regions, especially developing countries, still lack reliable data. This study respects the observed ordinal ranking of adaptation costs and effectiveness which, given the overwhelming uncertainty, can be considered as informative as a perfect replication of the data. Further details are described in Agrawala et al.

Table 6.1 Adaptation costs, adaptation effectiveness, and total climate change damages for a doubling of CO_2 concentration. Extrapolation from the literature and calibrated values

	Estimated adaptation costs (% of GDP)	Estimated adaptation effectiveness (% of reduced damage)	Calibrated adaptation costs in AD-WITCH (% of GDP)	Calibrated adaptation effectiveness in AD-WITCH (% of reduced damage)	Residual damages in AD-WITCH (% of GDP)	Total damage in AD-WITCH (% of GDP)	Total damages in Nordhaus and Boyer (2000) (% of GDP)	Total damages in the WITCH model (% of GDP)
USA	0.09	0.18	0.1	0.22	0.4	0.5	0.45	0.41
WEURO	0.18	0.13	0.27	0.13	1.63	1.95	2.84	2.79
EEURO	0.37	0.3	0.18	0.27	0.72	0.9	0.7	–0.34
KOSAU	0.48	0.16	0.19	0.18	0.81	0.98	–0.39	0.12
CAJANZ	0.09	0.2	0.06	0.11	0.14	0.25	0.51	0.12
TE	0.28	0.12	0.15	0.12	0.55	0.67	–0.66	–0.34
MENA	1.06	0.34	0.81	0.46	1.99	2.8	1.95	1.78
SSA	0.7	0.21	0.62	0.19	3.58	4.23	3.9	4.17
SASIA	0.49	0.19	0.68	0.23	3.72	4.38	4.93	4.17
CHINA	0.2	0.15	0.11	0.21	0.49	0.56	0.23	0.22
EASIA	0.4	0.18	0.45	0.21	1.75	2.2	1.81	2.16
LACA	0.13	0.38	0.24	0.25	0.96	1.24	2.43	2.16

(2011). Despite the effort made to gather new information, the AD-WITCH representation of climate change impacts still has some limitations. The description of non-market damages is only partial and AD-WITCH, like most IAM, abstracts from very rapid warming and large-scale changes of the climate system (system surprises).

6.3 COST-BENEFIT ANALYSIS OF THE OPTIMAL MIX BETWEEN ADAPTATION AND MITIGATION

AD-WITCH can be solved in two alternative settings. As a non-cooperative game, in which the 12 model regions behave strategically with respect to all major economic decision variables, including adaptation and emission abatement levels, by playing a non-cooperative game (the BaU scenario). This yields a Nash equilibrium, which does not internalise the environmental externality. The cooperative solution describes a first-best world, in which all externalities are internalised because a benevolent social planner maximises a global welfare function.[2] The latter solution concept is exploited in this section to characterise the optimal relationship between adaptation and mitigation.

A first important result of the analysis is that both mitigation and adaptation are part of the optimal response to climate change damages. Figure 6.1 plots the optimal expenditure on adaptation over time on the left axis and the optimal emission reduction path on the right axis. This is expressed as percentage emission reduction compared to a baseline in which countries do not cooperate on climate change mitigation.

Since adaptation addresses current and near-term damages, it remains low

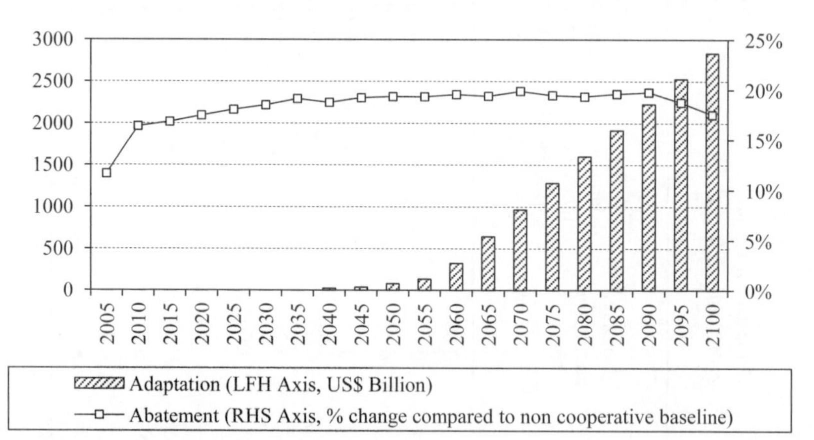

Figure 6.1 Mitigation and adaptation in the optimal climate change strategy

during the first two decades and becomes detectable in 2035 (USD2 billion). Afterwards, it increases rapidly reaching USD326 billion in 2060, peaking to nearly USD3 trillion in 2100. Optimal abatement leads to global emission reduction ranging between 15 percent and 19 percent compared to the BaU throughout the century, leading to a temperature increase of about 3–3.4°C above pre-industrial levels. Overall, the optimal climate change strategy would imply a 44 percent damage reduction, 78 percent of which is accomplished by adaptation and only the remaining 22 percent by mitigation. This is an efficiency first-best result.

Adaptation and mitigation are strategic complements in their contribution to damage and temperature reduction, as illustrated in Figure 6.2.

The Figure 6.2a shows that optimal adaptation alone could reduce residual

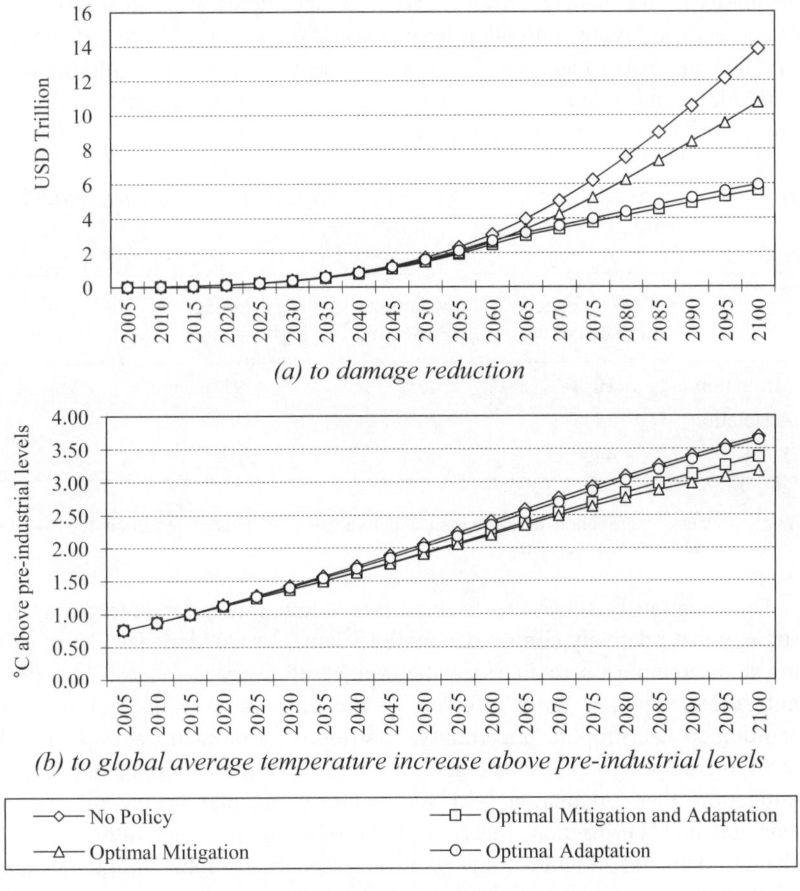

Figure 6.2 Contribution of adaptation and mitigation

damages up to 55 percent in 2100, averting about USD8 trillion of damages, whereas optimal mitigation alone would lower damage up to 20 percent, avoiding about USD3 trillion worth of damages. Although adaptation effectively deals with unavoidable damage, it has nearly no effect on temperature (Figure 6.2b). Mitigation is more effective at keeping global warming below dangerous thresholds and it is the only viable strategy to avoid catastrophic, potentially irreversible and inadaptable damages that would be triggered by higher temperatures. Hence, a climate strategy that aims at addressing both residual damages and the risk of future dangerous consequences should combine both strategies.

Discounting and the size of damages have major influences on the mitigation and adaptation mix because they govern the perception of present and future damages. Abatement is further increased when the discount rate decreases or the damage from climate change increases. The discounting effect tends to favour mitigation by increasing the weight of future damages. The damage effect increases future and present damages and calls for both mitigation and adaptation. Table 6.2 analyses the sensitivity of the mitigation-adaptation mix to damage and discounting.

*Table 6.2 Sensitivity analysis to climate change damages and to discount rate (*Low damage, high discount)*

	Low damage high discount	Low damage low discount	High damage high discount	High damage low discount
Mitigation	10%	40%	9%	43%
Adaptation	34%	20%	53%	30%
Total	44%	60%	62%	73%

Note: Relative contribution of adaptation and mitigation to damage reduction (2010–2100 cumulative undiscounted).

Unambiguously, when the discount rate is low, mitigation emerges as the strategy that relatively contributes more to damage reduction. The effect of a lower discounting is to increase the weight of future damages. Therefore mitigation, which is more effective in the distant#future, is preferred. In addition to catastrophic uncertainty, low discount rates are another factor justifying not only higher abatement, but also a relatively more intense use of mitigation than adaptation. On the contrary, adaptation prevails when damages and the discount rates are high. When present and future climate change damages double (high damage scenarios) both mitigation and adaptation efforts increase, but in relative terms, adaptation, which deals more effectively in the near-term, is preferred. This effect is strengthened by high discount rates. The maximising agent does not perceive larger damages that would prevail in the second part of the century, and mitigation is reduced in favour of adaptation because of the longer time distance between expenditure and returns.

The question that is addressed in the next section is whether this result is also robust to more ambitious stabilisation targets. In fact, it should be pointed out that the analysis presented in this section, even in the case of high damage and low discount rate, did not consider the risk of discontinuities and large scale, catastrophic events. AD-WITCH only represents changes in average conditions and in this setting aggressive mitigation appears to be less appropriate. Because low probability and high consequence events are long-term and highly uncertain, the effect of mitigation on damage reduction cannot be explicitly shown. In addition, the mitigation effort that would be required to deal with them is exposed to the subjective assumptions in policy decision making, such as the discount rate. As a consequence, a precautionary approach can be used to justify stronger mitigation actions capable of limiting long-term temperature increase below 2 or 2.5°C. Having set the long-term, the climate objective AD-WITCH can provide information on the most cost-effective mode of achieving that target.

6.4 COST-EFFECTIVE ANALYSIS OF THE OPTIMAL MIX BETWEEN ADAPTATION AND MITIGATION

If the damages considered do not include the risk of large-scale, catastrophic events and there is a stronger preference for the present, adaptation is the preferred option and only moderate mitigation is appropriate. When the temperature (or GHG concentrations) target becomes more ambitious, mitigation is more effective, as shown in the previous section. Thus, the policy question is whether adaptation is still justified, even if an increasing amount of resources are being allocated to finance the mitigation effort. What is the optimal mix between adaptation and mitigation when there is a climate policy that stabilises GHG concentrations at 550 CO_2-eq, leading to a temperature increase of 2.5°C above pre-industrial levels?

In this section, countries cooperate on mitigation policy, but the adaptation policy responds to cost-benefit considerations made at the regional level, in a non-cooperative manner. Concerning the optimal timing of emissions reduction and adaptation, Figure 6.3 shows that, provided climate policy is credible so that countries anticipate it, mitigation starts immediately even though initial climate damage is very low. Anticipation is needed in order to deal with the inertia of the carbon cycle and the energy system. Mitigation

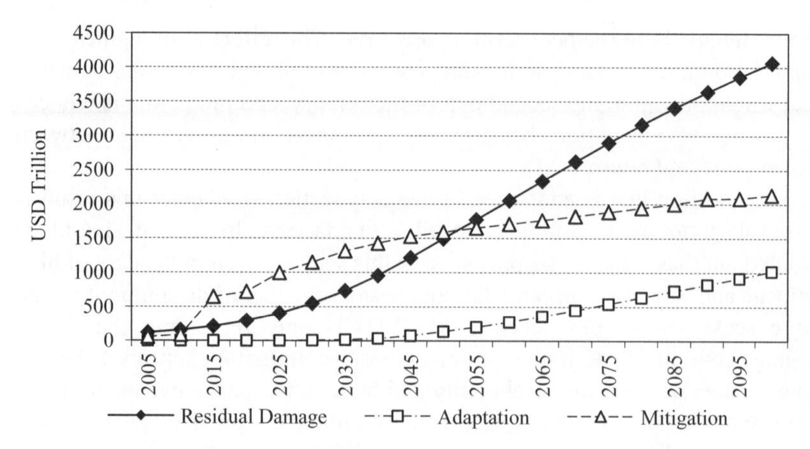

Figure 6.3 Adaptation, mitigation, and residual damage. Intertemporal path of the expenditure timing

options require substantial long-term investments to become competitive and deployed on a large scale and, therefore they must occur earlier. In contrast, adaptation expenditure increases with damage because it directly impacts climate change damage. As a consequence, it can be postponed until damage becomes significantly high.

Allocating resources to mitigation reduces what is left for other usages, such as adaptation, which in fact is halved compared to a situation in which adaptation would be the only viable option to deal with climate change. More precisely, the mitigation policy considered in this section lowers the need to adapt and crowds out adaptation expenditure. The crowding-out is particularly prominent after mid-century, when it reaches about 50 percent. Nonetheless, adaptation remains substantial and it still exceeds USD1 trillion in 2100.

To conclude, although a fairly ambitious mitigation policy target is adopted internationally and mitigation reduces climate damages, there is still room for adaptation. The net effect of combining adaptation and mitigation is a welfare improvement in the long-term. Initially, the additional expenditure on adaptation and the increased costs of mitigation are not compensated by the reduced damage, but as long as climate related damages increase, adaptation becomes more useful.

6.4 REGIONAL CONSIDERATIONS

Due to the local nature of adaptation and the differences in regional vulnerability, regional adaptation patterns may differ substantially from what the global picture suggests. Such diversity is shown in Figure 6.4, which emphasises the different size, timing, and composition of adaptive behaviour across developing and developed countries.[3]

Developing countries are more exposed to climatic damages, for at least two reasons: first, potential impacts are likely to be larger (70 percent of damages occur here). Second, they face an adaptation deficit: adaptive capacity (institutions, access to information, natural resources, and education) is lower. As a consequence, they are forced to spend more than OECD regions in all forms of adaptation as a percentage of their Gross Domestic Product (GDP) as well as in absolute terms. In 2100, adaptation expenditure in non-OECD countries more than doubles that of OECD regions. Not surprisingly, adaptation effort is particularly large in more vulnerable regions, namely Sub-Saharan Africa (SSA), South-Asia (SASIA), Middle East and North Africa (MENA). The effective availability of resources to meet adaptation needs in developing regions is particularly concerning. In 2050, developing countries are expected to spend around USD200 Billion (already twice the current flow of official development assistance), but approximately USD1.6 Trillion in 2100. The annuitised cost of adaptation to climate change in non-OECD countries is approximately USD500 Billion (or 0.48 percent of their GDP) against USD200 Billion (or 0.22 percent of GDP)

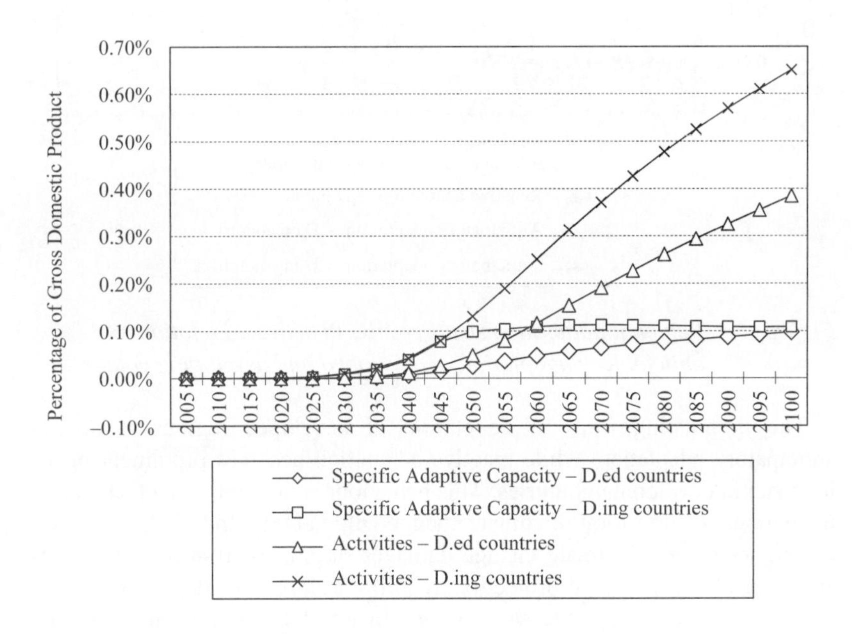

Figure 6.4 Regional adaptation strategy mix. Adaptive capacity building versus adaptation activities developed and developing countries

in OECD countries. In developing countries damage is not only higher, but also occurs earlier. In particular, non-OECD countries first need to build up a stock of adaptive capacity, an essential prerequisite for successful adaptation. Investments in specific adaptive capacity in developing countries are larger and grow faster during the first half of the century with respect those in developed countries.

Figure 6.5 shows how the composition of the adaptation portfolio also differs across countries.

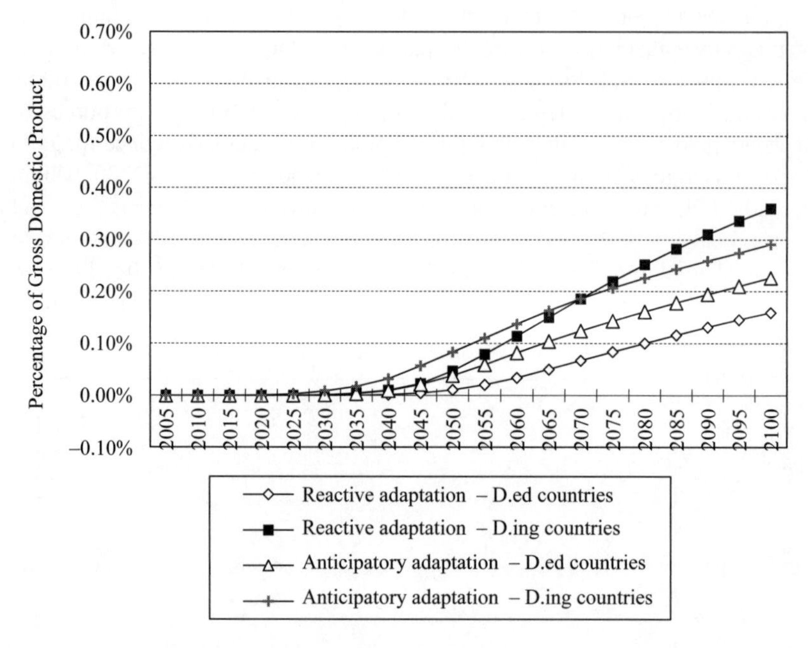

Figure 6.5 Regional adaptation strategy mix. Reactive adaptation and anticipatory adaptation in developed and developing countries

The dominating form of adaptation in developed regions is always anticipatory adaptation, while reactive adaptation becomes prominent in the long-run in developing countries. This behaviour is driven by two facts. First, the regional composition of climate change vulnerability. In OECD countries, the higher share of climate change damages originates from infrastructures and coastal areas. Their protection requires a form of adaptation that is largely anticipatory (of the stock type). In non-OECD countries a higher share of damages originates from agriculture, health, and energy sectors (space heating and cooling). These types of damages can be accommodated more effectively with reactive measures (of the flow type). Second, OECD countries are richer. They can then easily give up their present consumption to foster anticipatory adaptation, which is similar in nature to an investment as it will become productive in the future. On the contrary, non-OECD countries are compelled by resource scarcity to act in emergency.

The geographic differences are also important when considering adaptation in the presence of ambitious mitigation. When there is a global policy that stabilises GHG concentration at 550 ppme, OECD regions experience lower damages compared to what they would suffer under the sole implementation of optimal domestic adaptation. In non-OECD regions the opposite occurs: residual damages are higher under the mitigation policy than under optimal domestic adaptation. As discussed above, adaptation is more effective at reducing residual damage than mitigation, see Figure 6.2. As a consequence, mitigation reduces the need to adapt by a lower margin.

6.5 CONCLUSIONS

This chapter presents the extension of the WITCH model to include adaptation as a strategy that can be optimally combined in an integrated policy framework. The cost-benefit analysis performed with this set-up confirms that the joint implementation of mitigation and adaptation is welfare improving. It also shows that in a world without catastrophic events and where the decision maker has a strong preference for the present, adaptation is unambiguously the preferred option. However, even if a more aggressive mitigation policy were prescribed, advocating the precautionary principle approach, then adaptation would still be necessary, especially in developing countries. Reverting back to the question raised in the introduction about the optimal policy mix, the analysis carried out with the integrated assessment model AD-WITCH indicates that in the short-run, the optimal allocation of resources between adaptation and mitigation should be tilted towards mitigation. Adaptation becomes increasingly important in the longer-run. Therefore, if the aim is to reduce the probability of catastrophic and possibly irreversible climate related damages, aggressive mitigation actions need to be implemented soon.

The main challenge of short-term cooperation on adaptation should be to make developing countries less vulnerable and more resilient. This means that short-term cooperation on adaptation should mostly address capacity building (education, access to information, institutional capacity building) and soft adaptation measures including land use planning, integrated watershed management, adaptation technologies, and R&D.

Adaptive capacity, along with the potential impacts from climate change, determines the vulnerability of the system and is therefore a key priority.

Failure to build it would seriously undermine the effectiveness of any other adaptation activities. On the other hand, reactive adaptation becomes the main option in the long-run, after 2080, when expected damages would be larger. Technological innovation is expected to be particularly important because the largest potential for its application is to health and agriculture. These two sectors are particularly vulnerable in developing countries either because they lack an appropriate infrastructure system (the first) or because the climatic conditions are already extreme (agriculture). However, the lack of innovative capacity in developing countries creates a mismatch between where innovation is carried out and where innovation would be needed the most. Thus, international research cooperation could bridge such gap between where innovation is produced and where it is required.

Although the geographic distribution of climate change impacts and consequently of adaptation needs cannot be modified, the equity-adverse impact of climate change can be addressed with proper financial and technology transfers. Cooperation on adaptation might act as the carrot that helps to change developing countries' attitude on international cooperation on mitigation as well. Both the role and impact of innovation for adaptation and the existence of policy spillovers between adaptation and mitigation are topics that deserve more research.

NOTES

1. This chapter is based on and reproduces content from the following papers: Agrawala et al. (2011), Bosello (2008), Bosello et al. (2010, 2013).
2. Both the AD-WITCH and the WITCH model, also feature technology externalities due to the presence of Learning-By-Researching and Learning-By-Doing effects. The cooperative scenario internalises all externalities. For more insights on the treatment of technical change in the WITCH model see Bosetti et al. (2009).
3. For developing or non-OECD countries we refer to the following regions: CHINA – China and Taiwan, SASIA – South Asia, SSA – Sub-Saharan Africa, LACA – Latin America, Mexico, and the Caribbean, TE – Transition Economies, EASIA – South East Asia, MENA – Middle-East and North Africa.

REFERENCES

Agrawala, S. and S. Fankhauser (2008), *Economics aspects of adaptation to climate change. Costs, benefits and policy instrument*, Paris: OECD.

Agrawala, S., F. Bosello, C. Carraro, E. De Cian, E. Lanzi, K. De Bruin and R. Dellink (2011), 'Plan or react? Analisys of adaptation costs and benefits using integrated assessment models', *Climate Change Economics*, **2**(3), 1–36.

Bosello, F. (2008), 'Adaptation, mitigation and green R&D to combat global climate change. Insights from an empirical integrated assessment exercise', CMCC Research Paper No. 20.

Bosello, F., C. Carraro and E. De Cian (2010), 'Climate policy and the optimal balance between mitigation, adaptation and unavoided damage', *Climate Change Economics*, **1**, 71–92.

Bosello, F., C. Carraro and E. De Cian (2013), 'Adaptation can help mitigation: An integrated approach to post-2012 climate policy?', *Environmental and Development Economics*, **18**, 270–290.

Bosetti, V., E. De Cian, A. Sgobbi and M. Tavoni (2009), 'The 2008 WITCH model: New model features and baseline', Nota di Lavoro 95.2009, Milan: Fondazione Eni Enrico Mattei.

de Bruin, K.C., R.B. Dellink and R.S.J. Tol (2009a), 'AD-DICE: An implementation of adaptation in the DICE model', *Climatic Change*, **95**, 63–81.

de Bruin, K.C., R.B. Dellink and S. Agrawala (2009b), 'Economic aspects of adaptation to climate change: Integrated assessment modelling of adaptation costs and benefits', OECD Environment Working Paper No. 6.

Hope, C.W., J. Anderson and P. Wenman (1993), 'Policy analysis of the greenhouse effect-an application of the page model', *Energy Policy*, **15**, 328–338.

IPCC (Intergovernmental Panel on Climate Change) (2007), *Climate Change 2007: Impacts, Adaptation and Vulnerability. Contribution of Working Group II to the Fourth Assessment Report of the Intergovernmental Panel on Climate Change* [Parry, M.L., O.F. Canziani, J.P. Palutikof, P.J. van der Linden and C.E. Hanson, (eds)], Cambridge, UK and New York, NY, USA: Cambridge University Press.

Nordhaus, W.D. and J. Boyer (2000), *Warming the World. Economic Models of Global Warming*, Cambridge, MA: The MIT Press.

Parry, M. (2009), 'Closing the loop between mitigation, impacts and adaptation', *Climatic Change*, **96**, 23–27.

7. A Focus on the Latest Developments in the Modelling of Mitigation Options

Thomas Longden and Fabio Sferra

7.1 INTRODUCTION

The work of modelling is never completed, as models can and should be improved as soon as better or new information concerning processes or data becomes available. This chapter[1] aims at informing the reader on changes, improvements and augmenting features that have been made to the WITCH model during the last two years. In particular, as the modelling of emission mitigation options is a central component of the WITCH model, these improvements mainly concern the mitigation/technological side of model. Reflecting this on-going initiative to refine the WITCH model, this chapter reviews a range of possible mitigation options which have been incorporated into different versions of the model. With the WITCH model having a focus on climate policy, many of the additional mitigation issues chosen for further review are those which involve a greater focus on the provision or use of energy within the economy. The structure of the energy sector is of great importance to the overall emission profile of an economy and the introduction of new technologies provides many of the potential emission reductions modelled within both the base and the extended versions of the WITCH model. With the model design aiming to track the major actions which impact mitigation (such as R&D expenditures, investments in carbon-free technologies and adaptation), the range of additional mitigation issues that can potentially be covered is broad. Following this, the range of mitigation issues currently developed for inclusion into the WITCH model is diverse and ranges from the early retirement of power plants to changes in the use of fuels within key non-energy sectors. In Section 7.2 there is a review of a range of technology-based mitigation options within the energy sector, such as the early retirement of power plants (Capital Vintaging), additional (or refined) representations of renewable energy options (Refinement of Renewable Technologies), and the capture/storage of carbon dioxide (Capturing Carbon). Section 7.3 investigates the changes to policy costs that occur with the addition of oil trade between regions. Section 7.4 then focuses on the inclusion of a light duty vehicle transport sector to allow for mobility demand and breakthrough technologies in personally owned light duty vehicles.

7.2 TECHNOLOGY PORTFOLIO IN THE ENERGY SECTOR

The additional abatement options reviewed in recent versions of the WITCH model and relevant to the energy sector technology portfolio include: the early retirement of carbon intensive energy generation technologies, refinements to the representation of renewable technologies and carbon capture and storage (CCS), as well as the potential for direct air capture of CO_2. This is the same order that these abatement options will be discussed within the following subsections.

Capital Vintaging

Capital vintaging has been introduced in the WITCH model to differentiate between the existing power plants in the electricity generation sector and new power plants that possess cutting edge technology. As the average technical lifetime of the power plants is relatively high (varying from 25 to 45 years depending on fuel technology), the characteristics of the existing power plants can be considerably different compared to the new technology options currently available. New technologies tend to be characterised by higher energy efficiency and a higher load factor (representing the number of hours per year that they operate at full capacity). With capital vintaging the residual technical lifetime of the installed power plants at the base year is taken into account to reflect some inertia in the transition towards a low-carbon world. In order to achieve this, a differentiation of depreciation rates allows for the average residual economic lifetime of the installed power plants in 2005. While new capital has a depreciation rate that is constant over time, existing capital depreciates according to the average residual economic lifetime for each region (and fuel type). To calculate the average residual economic lifetime for each fuel and each region, we have utilised the data provided by WEO (2008), IAEA, NETL Coal Plant Database, as well as Ecofys (2008).

As a result of the change in the depreciation rate of existing capital, the amount of investments in the electricity generation sector is lower in the short-term when compared to the previous version of the WITCH model. The change in global investments reflects the changes in the depreciation rate and hence tends to be greater in the areas where coal, gas, and nuclear technologies are more heavily used. As the depreciation for old capital is based on the residual technical lifetime of the installed power plants the model provides an improved replication of power plant generation capacity in the short-term. These residual technical lifetimes change between regions and across technologies and, accordingly, the changes in investments at the regional level have a similar diversity. For both a Business-as-Usual (BaU) and a 550 ppm GHG stabilisation scenario, the most significant decrease in investments in coal is in China followed by India and TE. With respect to investments in gas the most significant decrease is in the USA and MENA, followed closely by the EU. For nuclear the decrease in investments is most significant in the EU, the USA and CAJANZ. These changes in investments tend to exist between 2005 and 2025 as the majority of the modifications to the depreciation rates have been made to reflect technical lifetimes. An exception to this is the case of notable nuclear expansion in the late 20th and early 21st century in non-OECD nations, such as China, India, and East Asia, where technical lifetimes extend for a greater number of years – up to 35 or 45 years in some cases.

Refinement of Renewable Technologies

Wind

Initially the modelling of wind in WITCH was combined with solar and had a combined learning-by-doing effect. With the extension of the model to specifically model Centralised Solar Power (CSP), the modelling of wind power has been refined and incorporates a supply curve which maps the marginal costs of different capacity levels (net of learning costs which are modelled endogenously). Hoogwijk et al. (2007) explored the dynamic change of electricity production cost with increasing penetration levels in the USA and OECD Europe. Within this journal article, the cost of wind electricity is dependent on four components: depletion and learning, spinning reserve, backup capacity, and discarded electricity. The depletion effect increases the cost of electricity due to the declining quality of the resource in terms of power, while there is a learning effect, which decreases the cost of the technology. The WITCH model has modelled this learning effect with the application of a learning curve with endogenously determined investments. The incorporation of a supply curve reflects the other components of the total cost for wind power. A fitted polynomial curve (shown in Figure 7.1) replicates the relationship provided by Hoogwijk et al. (2007). In addition, a lower bound on the capacity of wind power installed from 2010 onward has been set to ensure that capacity in 2010 is equal to the levels described within WEO (2010). The maximum technical potential of wind (EJ per year) for each region has been applied with the application of estimates from Ecofys (2008a).

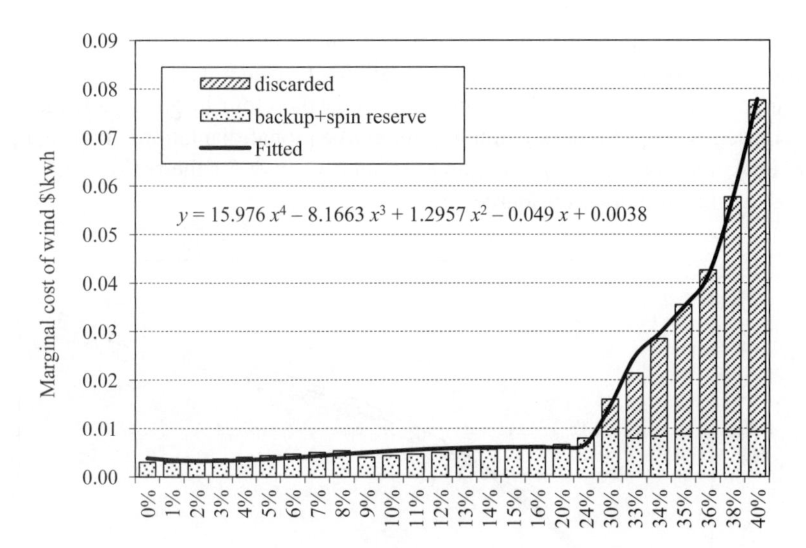

Note: Wind electricity production as a percentage of electricity production

Figure 7.1 Overall marginal cost of wind in US (excluding depletion and learning)

Biomass

Wind, hydro, and solar are not the only renewable technologies that demand interest with respect to climate change policy. Advanced production processes related to biomass conversion are of interest to integrated assessment models, such as the WITCH model, as they are a fuel source for use during biomass co-firing with coal in Integrated Gasification Combined Cycle (IGCC) power plants. In addition, biomass can be used in the production of biofuels for use in the transport sector. It is with these uses in mind that supply curves for woody biomass have been introduced into the WITCH model to define the costs of this input. The supply curves incorporated into the model represent the cost of woody biomass sourced from conventional plantations and short rotation forests for each region. These supply curves have been sourced from the Global Biomass Optimisation Model (GLOBIOM) by harmonising the model with parameters from the WITCH model to determine the supply function for woody biomass with respect to competing land use possibilities; such as managed forests, short rotation tree plantations, and cropland. With wood and food demand being determined by Gross Domestic Product (GDP) and population changes, regional estimates are produced by GLOBIOM with an allowance for up to 37 different crops and a minimum per capita calorie intake (Havlík et al., 2011). Given the land use change restrictions imposed in GLOBIOM, short rotation forests are not very sensitive to changes in prices as only cropland and grassland can be converted to plantations. Considering a given year, the change that is observed in the total supply is therefore in great part due to increased woody biomass coming from conventional plantations. Figure 7.2 reviews the global supply curve and the supply curve for the region with the most abundant biomass resources.

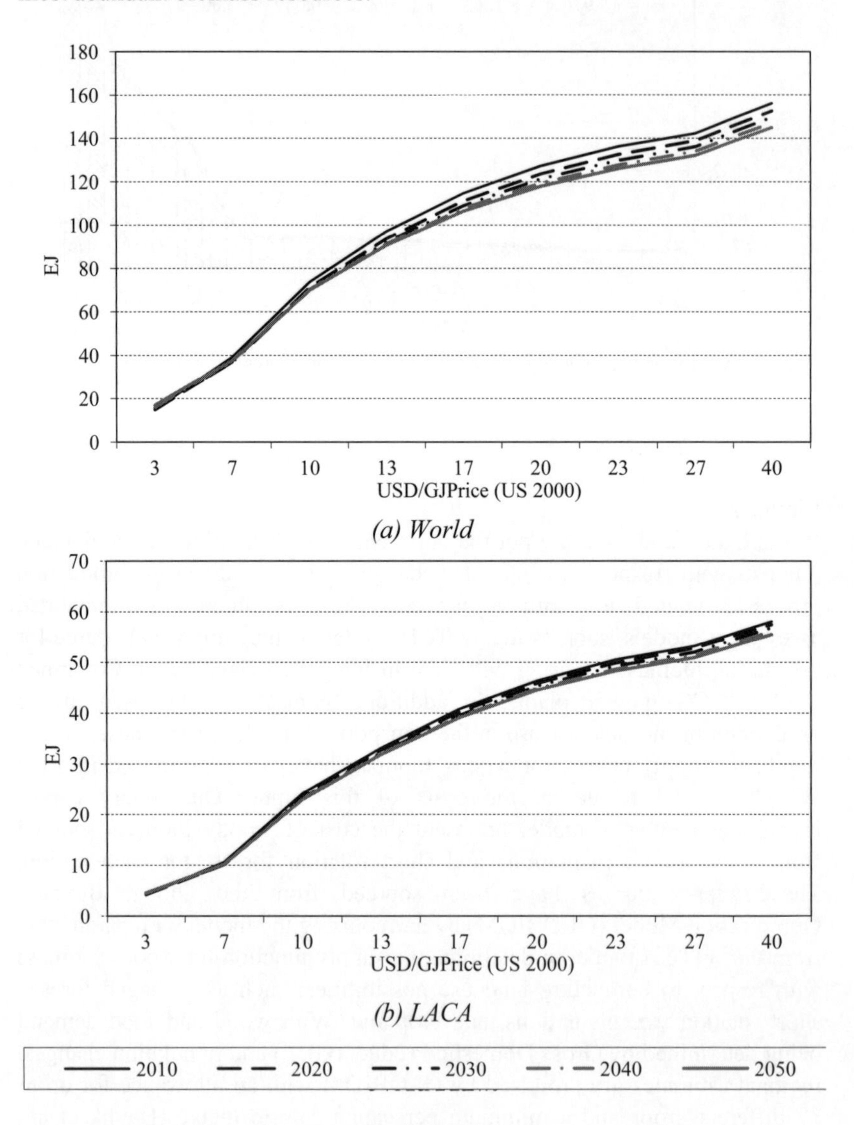

Figure 7.2 Supply curves from GLOBIOM

Correctly accounting for greenhouse gas emissions is a crucial element in a biomass power production analysis. A common assumption asserts that woody biomass is carbon neutral, in the sense that emissions resulting from its combustion have previously been compensated for by forest carbon sequestration. If bioenergy plantations were to encroach upon natural forests, major land use emissions occurring at this phase would compromise woody biomass carbon neutrality. Accordingly, the curves derived by GLOBIOM have been constructed using land use change restrictions that guarantee carbon neutrality. As a result, woody biomass is treated in WITCH as a carbon neutral energy. Notwithstanding this, fertiliser use, farming management, harvesting, and transportation imply additional emissions resulting from fossil fuel usage. A recent study (Evans et al., 2010) provides a survey on life cycle assessment for electricity production using woody biomass energy sources. While Evans et al. (2010) covers a large number of different electricity generation technologies it leaves out the technology considered in WITCH, for instance co-combustion with coal. As a result, the study by Dubuisson and Sintzoff (1998) is utilised as it provides final carbon emission factors for short-run coppices used to produce electricity with this type of technology. The stoichiometric coefficient (Mton/EJ) for woody biomass was computed by dividing the carbon content of woody biomass (Million tonnes of C) by the energy supply per unit of biomass (EJ). Life cycle emissions follow Dubuisson and Sintzoff (1998) with CO_2 emissions set at 12 Kg/GJ. The maximum biomass potential is 150 EJ globally, which corresponds to 3.75 GtC (1 EJ = 0.026 GtC).

Concentrated solar power

Super-grids powered by Concentrated Solar Power (CSP) have been included within the WITCH model as they provide a renewable source of power that when coupled with heat storage is able to produce power that may even be used for base-load power (hence it partially overcomes intermittency problems). The existing regional profile of the WITCH model has been applied to model long-distance transmission of CSP electricity within the USA, China and also between MENA and the European Union.

The type of CSP modelled is the production of solar thermal power using parabolic trough power plants. Such power plants are characterised by arrays of parabolic reflectors that concentrate solar radiation on to an absorber and convert it into thermal energy, which is then used to generate steam for a turbine. Power production with this kind of technology is strongly influenced by solar irradiance and atmospheric conditions. As solar thermal power employs direct sunlight, it is best positioned in areas without large amounts of humidity, fumes or dust that may deviate sunbeams; such as deserts, steppe or savannas (Richter et al., 2009). High Voltage Direct Current

(HVDC) cables, that are characterised by relatively low losses, allow to connect distant generation plants to demand centres. As a result, the focus of the modelling within this version of the WITCH model focuses on desert areas with high values of Direct Normal Irradiation such as those found in the MENA region, the north of China, and the South-West of the United States (Richter et al., 2009; Trieb, 2009; IEA, 2010). HDVC cables and the associated conversion stations are costly and in order for a Super-Grid to be installed there is a need for a significant and stable demand for energy, though from our simulations this does not seem to be a major problem, at least for the modelling of this technology in a macroeconomic model such as WITCH. Moreover, the case of trade between MENA and Europe presents additional complex considerations of supply security and geopolitical issues.

The amount of CSP electricity supplied to the grid of each region is determined through the combination of: (i) the generation capacity accumulated in each region, (ii) CSP plants operation and maintenance costs, (iii) the capacity of the Super Grid to transmit electricity from remote areas to the local grid; and (iv) operation and maintenance costs for the Super Grid. In accordance, the production function of CSP electricity is represented by a Leontief function. Power generation capacity in CSP accumulates through investments in concentrated solar power plants subject to a CSP capital depreciation rate and the unit investment cost of installing CSP generation capacity. These investment costs follow a one-factor learning curve depending on the cumulative world capacity of CSP power plants and, consequently, decrease as experience increases. To take into account the limited expansion possibilities at each time step – due to supply restrictions on intermediate goods – unit costs also increase with investments in the same period and region.

The production function for exported CSP electricity differs from the production function of CSP electricity consumed domestically due to different grid requirements and an additional index to represent exports. Investments in CSP generation and in the Super-Grid infrastructure enter the budget constraint together with O&M costs. The modelling of CSP is based on parabolic trough power plant technology, with nominal capacity of 50MW each, 100 percent solar share and equipped with integrated thermal storage units for seven hours (Kaltschmitt et al., 2007). The overall investment costs for such power plants are estimated at 260 million euro, while the operation and maintenance costs amount to approximately 5.1 million euro per year. The data refer to state-of-the-art technology and to installations in a geographic area with a high level of direct radiation (Kaltschmitt et al., 2007). The resulting simulations suggest that an extensive use of CSP both for domestic consumption or export will only occur in the second half of the century.[2] The introduction of the CSP option allows stabilisation policy costs to be reduced, and while policy is still costly (in terms of GDP loss compared to a business-as-usual scenario), the addition of CSP significantly decreases such losses in areas with high solar irradiance (such as MENA, China and the US) or in areas that can import this type of electricity (in our simulations, Europe). This high stabilisation-cost option value of CSP is particularly true for coal-intensive countries, and it reduces the option value of nuclear power and IGCC coal power with CCS. Regarding the opportunities of export of this zero carbon option, our simulations show that the gains for MENA, that is one of the major oil exporting regions, grow as further restrictions are imposed on power generation at a global level (in terms of climate policy or limitations on the expansion of other non-CSP generation technologies) and this may contribute to start to change the attitude of these countries towards global climate agreements.

Capturing Carbon

One technology that has received particular attention in the recent past is carbon capture and sequestration (CCS). CCS is a promising technology but is still far from large-scale deployment. Costs increase exponentially with the capacity accumulated by this technology. Having introduced a supply of woody biomass as a possible feedstock within the WITCH model, the capacity for its use within co-fired coal IGCC power plants has been established as an additional CCS option. Globally, experience with biomass (or waste) co-firing with coal covers about 150 power plants which are either at the pilot test stage or are used for commercial use (IEA Bioenergy, 2007). Biomass co-firing with coal requires relatively small changes at power plants already equipped with CCS technology and it also ensures some level of fuel flexibility. Currently biomass is co-fired in coal-fired power plants with a 10 percent fuel share (on energy basis) (NETBIOCOF, 2006) and there are only few examples with higher shares reaching 20 percent. Since IGCC power plants are equipped with the technology for capturing and storing the emissions produced from the combustion of both coal and biomass, biomass energy with CCS may yield negative net emissions. In accordance, woody biomass with CCS tends to be a key technology within the portfolio of CCS technologies within stabilisation scenarios.

The possibilities for capturing carbon may not be limited to co-firing with waste or biomass. The development of Direct Air Capture (DAC) envisions the development of a technology, which absorbs CO_2 by passing air with a high concentration of CO_2 over sodium hydroxide within a structure similar to a cooling-tower (Bickel and Lane, 2009; APS, 2011). In comparison to carbon removal from flue gas, DAC allows the process of carbon capture to employ economies of scale as it can be located amongst existing power generating systems (Keith et al., 2006). Since DAC can limit the cost of a stringent climate policy scenario and decreases the need for near-term mitigation, it can be an important additional abatement option in the middle of the century when the marginal abatement cost notably increases. This is the case under a 450 ppm GHG concentration target aimed at keeping the global temperature increase at 2 °C. Accordingly the WITCH model has been modified to review the impacts of implementing DAC and its impact on marginal abatement cost and energy use patterns. Having built a benchmark system based on current technological capabilities, APS (2011) released a report on the cost of DAC and its potential role within climate mitigation. The report concludes that while there is potential for the capture of decentralised CO_2 emissions,

> at least for the next few decades, unless there are dramatic cost reductions, direct air capture can be expected to be substantially more expensive than many other currently available options. (APS, 2011)

In applying the cost assessment developed within APS (2011), the WITCH model reviews both the relative cost of DAC in comparison to other CSS options and its effectiveness given the technology's need for electricity and the storage of captured carbon.

With no changes made to reflect technological improvements in DAC, the WITCH model applies the non-energy costs (such as capital costs and non-energy operating costs) and the electricity/non-electricity consumption levels sourced from the APS (2011) study for the whole century. With energy prices set to those derived by the WITCH model, the total cost of the technology allows for competing sources of electricity and the trade-off between supplementing and complementing existing CCS technologies. Note that an upper limit on the penetration rate of DAC has been set to 50 percent of the total carbon captured by other CCS technologies in the previous five-year period. DAC, like other forms of CO_2 absorption, consumes both electricity and non-electric energy. Electricity is used to power the fans and facilities associated with passing air over the sodium hydroxide to produce a cross-reaction with carbon hydroxide to form calcium carbonate. Non-electric energy is used for the heating of the calcium carbonate within a natural gas fuelled kiln to capture the released CO_2, which is then stored with the sequestered carbon from other CCS technologies. Reflecting the issue of net-carbon reductions, the WITCH model sources the electricity used to fuel DAC from the existing low-carbon or zero-carbon energies within the model (such as nuclear, renewables and carbon-intensive energy coupled with CCS facilities).

As shown in Figure 7.3 woody biomass CCS accounts for the largest share of carbon sequestration in a 450 ppm scenario while DAC becomes commercially viable in the middle of the century. DAC's share of carbon sequestration increases to over 60 percent of the total amount of carbon

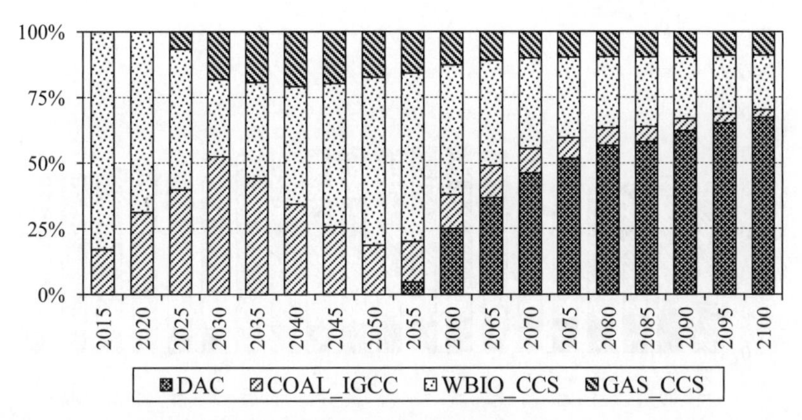

Figure 7.3 Composition of carbon sequestration

sequestration in the late part of the century. By capturing several Gt of CO_2 each year (rising from approximately eight Gt in 2070 to approximately 17 Gt in 2010), the emergence of DAC results in a decrease in the carbon price of about 73 percent at the end of the century, down from USD4000/Gt CO_2 to USD1000/Gt CO_2. The total policy cost (calculated using the percentage loss of GDP in comparison to the Business-as-Usual scenario) starts to reduce in comparison to the standard WITCH 450 ppm scenario and this corresponds with DAC coming online in 2055. A sharp reduction in the policy cost can be explained by a higher prevalence of the cheaper carbon-intensive energy sources, in comparison to the standard 450 ppm stabilisation scenario. Upon comparing the composition of world energy consumption, as in in Figure 7.4, the total energy consumption increases by more than 300 EJ at the end of the century. The most prominent difference is the consumption of oil, which in 2100 is eight times more than the case without DAC, while investments in nuclear and renewables are also increased due to the decreased demand for CCS technologies.

Without the deployment of DAC, the achievement of a 450 ppm target within the WITCH model is difficult when the carbon market is incomplete or when the largest emitters are absent from the abatement agreement. While the carbon price is higher in a scenario without the contributions of India and China (reaching almost USD1500/Gt CO_2), the potential importance of DAC in the long-term is reflected in the achievement of GHG concentrations at the 450 ppm level. The results of incorporating DAC into the WITCH model are similar to those of the APS (2011) report as the analysis in both cases show that the DAC deployment should occur in parallel with other CCS technologies, that its deployment will occur slowly and is likely to occur in the latter half of the century under a strong policy reaction to climate change (such as a 450 ppm stabilisation target).

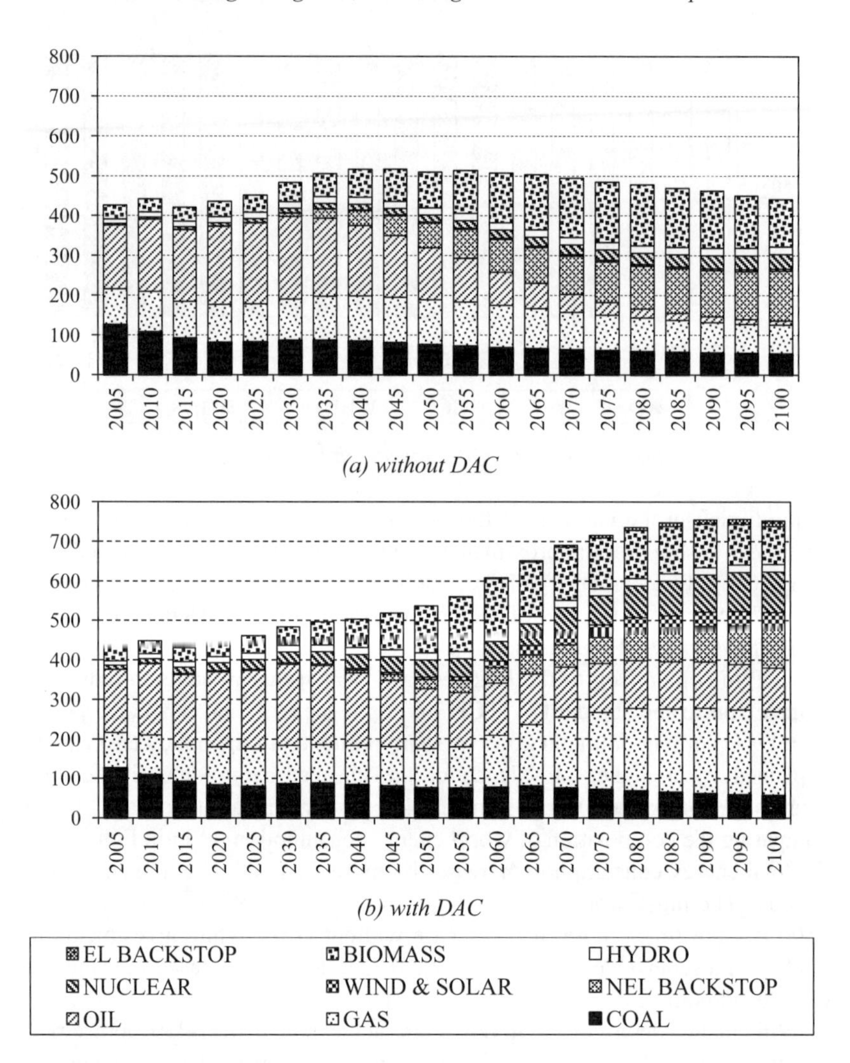

Figure 7.4 Composition of energy consumption

7.3 INVESTMENTS IN OIL UPSTREAM

The emergence of a low-carbon world (such as that implied by either GHG stabilisation scenario) would mark the end of an 'oil age' and provide widely unexplored economic, technological, and geopolitical implications. It is with this in mind that the WITCH model has been extended to incorporate an oil sector that evolves endogenously across all regions. This oil sector includes up to eight categories of oil, reflecting extraction costs and emissions related to oil extraction for each category. The production of oil is a function of extraction capacity built through endogenously determined investments. The cost of additional oil extraction is also endogenous and depends both on a short-term component, which mimics cost spikes when expansion capacity grows too fast, and on a long-term component, which reflects oil scarcity. Thus, the total expenditure in the oil sector is also endogenous. Once extracted, oil can be used for domestic consumption or it can be traded internationally. The price of oil emerges endogenously as an outcome of a Nash game among the all of the WITCH regions.

The model is calibrated to replicate base year oil production as well as imports and exports. We assume that oil traded internationally is homogeneous and therefore we have a unique international oil price. The cost of additional oil capacity is region-specific and accounts for both long-term exhaustibility and for short-term frictions that might arise in the supply chain when too much capacity is installed in a short-time period. Crude oil is used both in the electric and in the non-electric sector in WITCH. Oil demand is covered by means of domestic production of each category of oil and/or by means of net oil imports from the international oil market. In oil-exporting regions, domestic production of oil is greater than domestic consumption and net imports are negative. Oil production in a given year cannot exceed the extraction capacity cumulatively built in the country. Equilibrium in the international oil market requires that excess demand of oil is equal to zero at any given time period. Oil is valued at international market prices for regions and a mark-up is added to account for local factors that affect the cost of oil for final users and this mark-up can be greater or lower than zero. Investments for oil extraction are equal to the expenditure for financing the expansion of oil capacity and are both region and time specific. We assume that labour is not necessary to extract oil. This is a simplification that does not bear relevant implications since the oil-extraction sector is highly capital intensive.

Oil resources are derived from Rogner (1997) and they are assumed not to grow over time. Oil resources can be separated into eight categories (or less). In Table 7.1 they are aggregated into two categories: conventional oil (categories I–IV) and non-conventional oil (categories V–VIII). In 2005, non-conventional oil production is negligible and concentrated only in a few regions: Canada (CAJANZ aggregation), Brazil and Venezuela (LACA aggregation), and the USA. Oil imports and exports in the base year are calibrated using data provided by Enerdata (World Energy Database). In 2005, the USA is the largest oil importer (4.83 Billion Barrels per year), followed by WEURO, CAJANZ and CHINA. The largest oil exporter is MENA (7.6 Billion Barrels per year) followed by TE, LACA and SSA in decreasing order.

Table 7.1 Oil overview in 2005

	USA	WEURO	EEURO	KOSAU	CAJANZ	TE	MENA	SSA	SASIA	CHINA	EASIA	LACA
Resources (Bln. Barrels)												
Conventional (Cat I-IV)	347	120	13	15	59	538	1341	130	20	186	76	445
Non Conventional (Cat. V–VIII)	1749	416	39	989	1823	1336	3007	331	32	1737	269	3848
Production (Bln. Barrels)												
Conventional (Cat I–IV)	2.1	1.8	0.0	0.2	1.0	4.1	10.1	1.9	0.3	1.3	0.9	3.9
Non Conventional (Cat. V–VIII)	0.0	0.0	0.0	0.0	0.1	0.0	0.0	0.0	0.0	0.0	0.0	0.1
Net Exports (Bln. Barrels)	–4.8	–2.8	–0.5	–1.0	–1.5	2.4	7.6	1.6	–0.8	–1.4	–0.6	1.7
Consumption (Bln. Barrels)	7.0	4.6	0.5	1.2	2.6	1.7	2.5	0.3	1.1	2.7	1.5	2.3

Under a BaU scenario, world consumption of oil doubles during the century (in comparison to consumption in 2005), while the volume of oil traded internationally increases by only 60 percent. This is explained by the exploitation of vast non-conventional oil resources in countries with high domestic demand. The value of oil traded internationally, continues to increase due to the growing price of oil. The stabilisation scenario is constructed assuming that all regions agree on a global trajectory of emissions to stabilise GHG concentrations in the atmosphere at 550 ppm CO_2-eq at the end of the century. The regional outlook of the oil sector changes dramatically in the stabilisation scenario. Oil consumption drops substantially in all regions, with respect to the BaU scenario. With respect to 2005, aggregate consumption of Transition and Developing economies grows until 2025, while consumption in high-income economies peaks as early as 2015. The production of oil also changes substantially under climate policy. Non-conventional oil is extracted only in minimal quantity and MENA along with TE are able to supply all oil needed until the end of the century. Between 2040 and 2085 international oil trade is dominated by two macro-regions: the Middle East and North Africa and Transition Economies. The pattern of oil production triggered by the stabilisation policy entails that Western Europe, CHINA and the USA increase their reliance on foreign oil to supply their domestic consumption over the century. However with total demand of oil decreasing substantially, MENA, LACA and TE dominate a market that rapidly shrinks and this leads them to being net losers. In addition, their grip on the energy systems of Europe, the USA, and China vanishes under such stabilisation scenarios. At the global level, the cost of the stabilisation policy is quite sensitive to assumptions on the availability of the least cost non-conventional oil. The results of a sensitivity analysis within Massetti and Sferra (2010) reveal a need for an accurate description of fossil fuels extraction sectors to assess the macroeconomic consequences of a stabilisation policy. In conclusion, it should be noted that introducing a detailed description of the oil sector into the WITCH model results in an increase in policy costs and the regional distribution of these costs shifts towards oil-exporting regions. With climate policy expected to reshape the geo-politics in oil rich areas of the world, oil-rich countries will have a strong incentive to undermine climate agreements, unless they are adequately compensated for their losses.

7.4 DEMAND FOR MOBILITY AND TRANSPORT OPTIONS

With forecasts of transport demand in less developed and fast growing nations being approximately three times the rate of the OECD, considerable potential for growth in travel within even the most conservative economic scenarios is expected (Kahn Ribeiro et al., 2007). In order to analyse long-term trends in transport and their repercussions on the rest of the economy a transport module representing the use and profile of Light Duty Vehicles (LDVs) has been introduced into the WITCH model. LDVs have been selected as the vehicle type of interest as they have been identified as being one of the most favoured modes of transport and also one of the most damaging (Chapman, 2007). The addition of the transport module into the WITCH model allows for the evaluation of how the choice between LDVs will affect emissions as well as how these choices are likely to be impacted by climate change policies. The incorporation of the LDV transport sector has been conducted in a manner which allows for a range of emission mitigations to be possible, this includes: increased fuel efficiency, the introduction of alternative fuels and vehicle types, as well as curbed demand through decreases in the amount of kilometres travelled. For an extensive description of the modelling of these mitigation options, refer to Bosetti and Longden (2012).

Demand for vehicles has been set based on the assumption that constant travel patterns correspond to given levels and growth rates of GDP and population. The relationship between vehicle ownership and national income has been established within Dargay and Gately (1999) and applied to forecasting within WBCSD (2001). This assumption is important, as the demand for private transport will likely continue to be high and have a strong correlation with national income – unless a significant change in the provision of public transport occurs. With its current framework the model reviews the continuation of constant travel patterns and the constraint that this will place on the achievement of emissions reductions. This means that increased LDV travel (in terms of kilometres travelled per vehicle) as well as the costs of the vehicle and fuel expenditure directly impact utility through the corresponding effect of decreasing consumption on other goods and services. The model separates consumption in transport from the rest of consumption, which allows for the direct modelling of the costs involved in switching between vehicles and fuels for a given demand of mobility. Investments in vehicle capital and supplementary costs decrease the level of consumption. A Leontief function combines exogenous costs of vehicles with fuel costs and operation and maintenance (O&M) costs. Fuel costs depend upon the vehicle chosen and the price of fuel derived in the energy sector, where the fuel demand for oil, gas, biofuels and electricity directly compete with other electric and non-electric uses of fuels.

The modelling of transport within WITCH involves the distinction between the aggregate level of consumption and the level of consumption net of transport related expenditures with the ultimate budget constraint being set to consumption net of output, transport expenditure and investments. With the utility function defined, we can turn our attention to the modelling of the transport sector itself. Starting with the level of investments in vehicles in time period one, the subsequent period's capital stock of LDV is equal to the level of capital remaining after depreciation and the additional capital implied by investments undertaken at the prevailing investment cost of vehicles. With an exogenous estimate of the amount of mobility and hence the vehicle capital demanded in each region, a constraint is placed on the amount of capital in each period for each region. The amount of fuel demanded by each vehicle has been defined as a function of the average fuel efficiency of the vehicle for the amount of kilometres travelled per year using the different fuel technologies, *e*, and the amount of fuel efficiency improvements (FEIs) to date. The amount of fuel efficiency improvement is derived as a function of time (defined as the number of 5-year time spans that have passed) and a fuel efficiency factor. The average fuel efficiency variable has been set to the 2005 level for each vehicle type and applying different fuel efficiency factors produces different FEI curves. The fuel efficiency factor adopted in the base scenario has been set to intersect the US EIA forecast for fuel economy in the USA for the year 2030. The amount of fuel is defined using terawatt hours for direct comparability across fuel types and is linked to the existing WITCH model structure for each of the energy technology types. The range of vehicles, *LDV*, introduced into the model has been selected to provide a representative overview of the type of vehicles expected to come into contention for successful market penetration in the medium to long-term future. While each of these categories have different fuel economy and vehicle cost levels, the results are discussed in the following terms: Traditional Combustion Engine vehicles (TCEs), hybrid vehicles (HYBRIDs), biofuel vehicles (BIOFUELs), advanced biofuel vehicles (ADV BIOFUELs), Plug-in Hybrid Electric-drive Vehicles (PHEV) and Electric Drive Vehicles (EDV)s.

Figure 7.5 shows the cost of three key vehicle types across six different time periods for the USA and compares these to that vehicle's share of global sales within the LDV transport sector. (Note that the price of EDVs in the USA is representative of those in all other regions – except for the fuel and carbon cost components.) A decrease in the share of TCE vehicles within 2020 and 2030 is due to the use of biofuels as an alternative fuel source. In 2050 a further decrease occurs with the introduction of EDVs in the US, which are primarily fuelled by electricity from woody biomass with IGCC. With respect to this it should be noted that depending on the fuel source chosen for the corresponding electricity demand, fuel cost, and carbon cost can be very low. The introduction of EDVs within the USA in 2050 corresponds with higher vehicle costs, but is offset by considerable emissions reduction possibilities compared to the alternatives. Hence the USA tends to buck the trend of having HYBRID vehicles come in first (or simultaneously in the case of the Rest of the OECD) with the introduction of EDVs following after further cost reductions. After 2050 there is a gradual move towards battery fuelled technologies with EDVs dominating worldwide in 2100. Figure 7.5 shows that for the USA, the cost of carbon and the price of oil lead to the cost of employing TCE vehicles increasing in the latter half of the century. At the same time, the cost of an EDV is stabilising at a level below that of TCE vehicles from 2090 onwards.

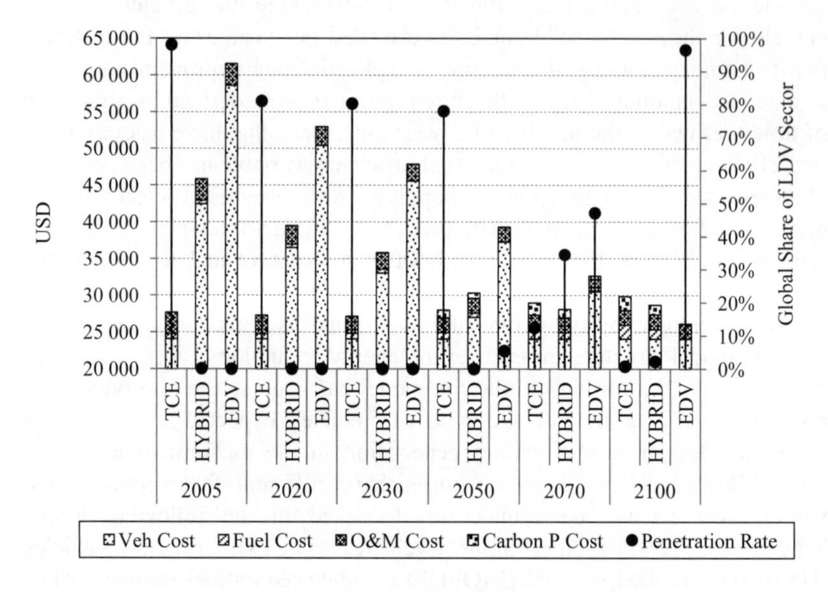

Figure 7.5 *Cost of new vehicle: Cost of a new vehicle in the USA and percentage share of global vehicles*

The sensitivity of the model to the existence of EDVs can be explained once we consider the amount of aggregate emissions from within the LDV transport sector for two additional stabilisation scenarios and compare the global carbon prices. Figure 7.6 highlights the sensitivity of the carbon price to higher aggregate emissions (attributed to a scenario with no EDVs – entitled the 'No EDV' scenario) and the insensitivity of the carbon price to lower aggregate emissions from within the LDV transport sector (attributed to a scenario with an earlier introduction of EDVs due to lower battery costs, entitled the 'Bat Cost' scenario). Within the 'No EDV' scenario the energy sector has had to compensate for continued reliance on fossil fuels and no decrease in the amount of travel performed or the amount of vehicles

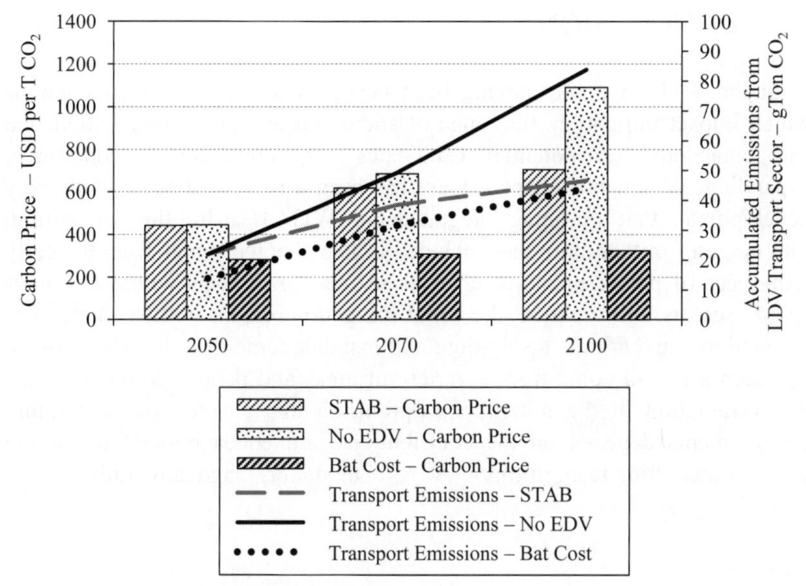

Figure 7.6 Carbon price comparisons

purchased. Indeed, a range of scenarios show a similar – trend over time and a stabilisation of world concentrations of GHGs at 550ppm CO_2-eq at a cost of approximately USD650 per tonne of carbon in 2100. The case of 'No EDV' diverges in the middle of the century, with the price of carbon steadily increasing until approximately USD1173 per tonne of carbon. This matches the trends in the global costs of climate policy (as a percentage of discounted GDP). In all scenarios, except for the 'No EDV' and 'Bat Cost' scenario, the World policy cost tends to be 2.6 percent of global Gross World Product (GWP). For the 'No EDV' scenario the policy cost is approximately 3.5 percent of global GDP. This confirms that the electrification and decarbonisation of the LDV transport sector is a notable issue with respect to the achievement of a cost effective climate policy. The case of improvements in battery costs result in policy costs of 2.2 percent of global GDP and reflects the relative insensitivity of the policy cost to lower emissions from the LDV transport sector. In addition to the carbon price, the cost of having no breakthrough in EDVs also impacts the world oil price. An increase in the oil price within the 'No EDV' scenario results in a mark up of approximately 50 percent of the STAB scenario price in 2100. Already significant, these costs are sure to rise in the complementary case of no electrification within freight transport, a sector of interest that is currently being directly modelled within WITCH.

7.5 CONCLUSION

With the WITCH model having been designed to track the major actions which impact mitigation, the range of additional mitigation issues modelled and considered as potential candidates for modelling is continually expanding. Within this chapter, there has been a review of some of the key developments that are being integrated into WITCH at the time of writing. The range of mitigation issues included in this chapter ranges from the early retirement of power plants to changes in the use of fuels within key non-energy sectors. In little time, the list of mitigation options will be added to as innovations, inventions, inspiration, and insights come to light. Their source has been and will come from a variety of areas and people. Reflecting this, the formulation of the mitigations options in this chapter and any future developments depends on the contributions and endeavours of all of the WITCH modelling team (http://www.witchmodel.org/pag/team.html).

NOTES

1. The formulation of this chapter would not have been possible without the contributions and endeavours of the expansive WITCH modelling team (link to the team page of the WITCH model website – http://www.witchmodel.org/pag/team.html) – in particular we should acknowledge the contributions of Valentina Bosetti, Carlo Carraro, Chen Chen, Enrica De Cian, Emanuele Massetti, Elena Ricci, Renato Rosa and Massimo Tavoni. Each section has been constructed using contributions from the researchers involved in developing the different model versions. In addition, section 7.1.2 utilises materials sourced from Massetti and Ricci (2011), section 7.2 utilises materials sourced from Massetti and Sferra (2010) and section 7.4 utilises materials sourced from Bosetti and Longden (2012).
2. For an extensive description of the modelling and of the results, please refer to Massetti and Ricci (2011).

REFERENCES

APS (2011), 'Direct air capture of CO_2 with chemicals: A technology assessment for the aps panel on public affair' available at: http://www.aps.org/policy/reports/popa-reports/loader.cfm?csModule=security/getfile&PageID=244407.

Bickel, J.E. and L. Lane (2009), 'Climate engineering' in B. Lomborg (ed.), *Smart Solutions to Climate Change: Comparing Costs and Benefits*, Cambridge: Cambridge University Press, pp. 9–51.

Bosetti, V. and T. Longden (2012), 'Light duty vehicle transportation and global climate policy: The importance of electric drive vehicles', Nota di Lavoro 11.2012, Milan: Fondazione Eni Enrico Mattei.

Chapman, L. (2007), 'Transport and climate change: A review', *Journal of Transport Geography*, **15**, 354–367.

Dargay, J. and D. Gately (1999), 'Income's effect on car and vehicle ownership, worldwide: 1960–2015', *Transportation Research Part A*, **33**, 101–138.

Dubuisson, X. and I. Sintzoff (1998), 'Energy and CO2 Balances in different power generation routes using wood fuel from short rotation coppice', *Biomass and Bioenergy*, **15**, 379–390.

Ecofys (2008), 'Efficiency and capture-readiness new fossil power plants', MVV Consulting, July 2008.

Ecofys (2008a), 'Global potential of renewable energy sources: A literature assessment – Background Report', ECOFYS 2008, Monique Hoogwijk, by order of: REN21 – Renewable Energy Policy Network for the 21st Century.

Enerdata (2007), 'World energy database'.

Evans, A., V. Strezov and T.J. Evans (2010), 'Sustainability considerations for electricity generation from biomass', *Renewable and Sustainable Energy Reviews*, **14**, 1419–1427.

Havlík, P., U.A. Schneider, E. Schmid, H. Böttcher, S. Fritz, R. Skalský, K. Aoki, S. De Cara, G. Kindermann, F. Kraxner, S. Leduc, I. McCallum, A. Mosnier, T. Sauer and M. Obersteiner (2011), 'Global land-use implications of first and second generation biofuel targets', *Energy Policy*, **39**(10), 5690–5702, doi:10.1016/j.enpol.2010.03.030.

Hoogwijk, M., D. van Vuurenb, B. de Vriesb and W. Turkenburga, (2007), 'Exploring the impact on cost and electricity production of high penetration levels of intermittent electricity in OECD Europe and the USA, results for wind energy' *Energy*, **32**, 1381–1402.

IAEA Power Reactor Information System – PRIS database. Available at: http://www.iaea.org/programmes/a2/index.html

IEA (2010), 'Energy Technology Perspectives 2010', Paris: International Energy Agency.

IEA Bioenergy (2007), Global Cofiring Database. Version 1.0 2005, IEA Bioenergy Task 32: Biomass combustion and cofiring. Available at: www.ieabcc.nl

Kahn Ribeiro, S., S. Kobayashi, M. Beuthe, J. Gasca, D. Greene, D.S. Lee, Y. Muromachi, P.J. Newton, S. Plotkin, D. Sperling, R. Wit and P.J. Zhou (2007), 'Transport and its infrastructure', in IPCC (Intergovernmental Panel on Climate Change), *Climate Change 2007: Mitigation. Contribution of Working Group III to the Fourth Assessment Report of the Intergovernmental Panel on Climate Change* [B. Metz, O.R. Davidson, P.R. Bosch, R. Dave and L.A. Meyer (eds)], Cambridge, UK and New York, NY, USA: Cambridge University Press.

Kaltschmitt, M., W. Streicher and A. Wiese (2007), *Renewable Energy*, Berlin: Springer-Verlag.

Keith, D., M. Ha-Duong and J. Stolaroff, (2006), 'Climate strategy with CO2 capture from the air', *Climatic Change*, **74**, 1–3.

Massetti, E. and E.C. Ricci (2011), 'Super-grids and concentrated solar power: A scenario analysis with the WITCH Model', Nota di Lavoro 47.2011, Milan: Fondazione Eni Enrico Mattei.

Massetti, E. and F. Sferra (2010), 'A numerical analysis of optimal extraction and trade of oil under climate policy', Nota di Lavoro 113.2010, Milan: Fondazione Eni Enrico Mattei.

NETBIOCOF (2006), 'D14: First state-of-the-art report', Integrated European Network for Biomass Co-firing (NETBIOCOF), available at: www.netbiocof.net

NETL (National Energy Technology Laboratory), 'Coal plant database', US DOE, available at: http://www.netl.doe.gov/energy-analyses/hold/technology.html

Richter, C., S. Teske and J.A. Nebrera (2009), 'Concentrating solar power global outlook 09', Greenpeace International/European Solar Thermal Electricity Association (ESTELA)/IEA SolarPACES, Report 2009.

Rogner, H.-H. (1997), 'An assessment of world hydrocarbon resources', *Annual Review of Energy and the Environment*, **22**, 217–262.
Trieb, F. (2009), 'Combined solar power and desalination plants: Techno-economic potential in Mediterranean partner countries', German Aerospace Center (DLR).
US EIA, 'Residential transportation historical data tables', Table 11: Fuel economy, available at: http://www.eia.doe.gov/emeu/rtecs/archive/arch_datatables/rtecshist_datatables.html
WEO (2008), *World Energy Outlook 2008*, Paris: OECD/IEA.
WEO (2010), *World Energy Outlook 2010*, Paris: OECD/IEA.
World Business Council for Sustainable Development Mobility (WBCSD) (2001), 'World mobility at the end of the twentieth century and its sustainability', available at: www.wbcsdmotability.org.

8. Conclusions

Emanuele Massetti and Massimo Tavoni

8.1 RECENT DEVELOPMENTS OF THE WITCH MODEL

Condensing the work of integrated assessment models into a book, as it has been done for the case of WITCH, creates an inevitable tension between old and new generations of the model. WITCH is a moving target, which develops and unfolds over time as new requests arrive from policy, new topics emerge as research relevant, new data and methods become available and new people with new skills join or leave the team. To partially capture the evolutionary nature of the model, this last section summarizes the most recent development of the models which have not made it into the book, but which are likely to play an important role in the result of the research being produced as we write. The reader can see these changes unfold live on the model website (www.witchmodel.org), but here we provide a short summary of the most important ones.

Climate

The climate module of WITCH has been completely revised, in order to account for the recent development in the Earth System Model community, and in order to provide an outcome which is able to reflect the deep uncertainties characterizing the carbon cycle and the overall climate.

To this end, WITCH is now coupled with the MAGICC6 model, a well-known reduced form climate model which has been calibrated on the Climate Modeling Intercomparison Exercises 4 and 5 ensembles (CMIP4 and CMIP5). We use the probabilistic version of MAGICC, which allows us to determine the full distribution of the temperature and climate outcome. Coupling WITCH and MAGICC is possible only when running WITCH in the non-cooperative setting; WITCH takes the emission budget as a constraint, and returns emissions profiles, which are then fed into MAGICC which returns the climate outcome. If needed, this process is repeated until the desired climate is achieved.

When running in a cost-benefit, full cooperation mode, MAGICC cannot be used, given the dynamic nature of WITCH. To fill this gap, the 3 box climate module described in this book has been recalibrated in order to yield results which are consistent with MAGICC.

An additional climate model, called SNEASY, has been coupled to WITCH. SNEASY has been calibrated on the latest CMIP5 model ensemble, and can be linked to WITCH both offline (as for MAGICC) or hard-linked through OBOE (but this increases the computational solution time).

Air Pollution

A specific module has been developed to keep track of air pollution and to analyze the important relation between climate and air pollution control policies.

To this end, a detailed sectorial and regional mapping between WITCH and the Representative Concentration Pathways (van Vuuren et al., 2011), EDGAR[1] and TIMER[2] data has been carried out. The most important substances (SO2, NOX, BC, OC, VOCs) are now tracked in the model, though for some sources specific and exogenous assumptions have been made. The climate impacts of these pollutants are fully accounted for in the model through MAGICC.

Several air pollution policies can be activated in the model, which account for various degrees and stringency of air pollution legislation. The model also generates emission control factors through a relation between air pollution abatement and income. Ongoing work includes the quantification of the damages of air pollution, so as to allow for a fully fledge cost benefit analysis.

Land Use

The coupling of WITCH with the land use cluster models developed at IIASA (GLOBIOM and G4M) has been further strengthened, thanks to the development of a large ensemble of model runs from the IIASA side, which are being incorporated into the WITCH model. These span a variety of combinations of different socio-economic assumptions and provide a coherent framework to assess bioenergy and forestry based mitigation options.

Renewable Energy Systems and their Integration

The description of the renewable energy systems (RES) has been considerable improved in WITCH, which now features a more detailed description of RES in terms of technologies (wind on and off shore, solar photovoltaic and concentrated solar power), a precise assessment of the resources base for each of them at the regional level, and more accurate description of the implications of integrating RES into the power system. This is modeled via a series of additional equations which account for the flexibility provided by each power technology to the electricity system, which differentiate capacity factors across technologies, and which capture the costs of RES integration when deployed at large scale. The model now also tracks investments in the power grid, and provides options to store electricity as an additional avenue for managing the integration of RES.

Solution

The model is solved in parallel, with each region being allocated a node, and via iterations. This allows making the most of the computational facilities at the CMCC (Centro Euro-Mediterraneo sui Cambiamenti Climatici) which have been upgraded and can now count on 7712 cores and 160 Tflops of computing power. This allows performing several thousand runs of WITCH per day, an important requirement for carrying out large model ensembles and for accounting for uncertainties.

Documentation

An effort is also under way to provide improved and extensive model documentation, for both internal and external purposes. The documentation provides a searchable interface which links the documentation of each module to the equations and to the actual lines of code within the model. All the documentation is accessible at the website www.witchmodel.org.

NOTES

1. See http://edgar.jrc.ec.europa.eu/index.php.
2. See http://www.pbl.nl/en/publications/2001/TheTargetsIMageEnergyRegionalTIMERModel TechnicalDocumentation

REFERENCE

van Vuuren, D.P., J.A. Edmonds, M. Kainuma, K. Riahi and J. Weyant (2011), 'A special issue on the RCPs', *Climatic Change*, **109** (1–2), 1–4.

9. Complete List of Publications that Use WITCH

This section provide references to all publications in peer-reviewed journals that use the WITCH model. The list was updated in November 2013. The interested reader will find more recent publications listed at the website www.witchmodel.org

2014 (Forthcoming)

Bowen, A., E. Campiglio and M. Tavoni (2014), 'A macroeconomic perspective on climate change mitigation: Meeting the financing challenge', forthcoming Special Issue *Climate Change Economics*.

Calvin, K., M. Wise, D. Klein, D. McCollum, M. Tavoni, B. Van Der Zwaan and D. Van Vuuren (2014), 'A multi-model analysis of the regional and sectoral roles of bioenergy in near- and long-term CO_2 emissions reduction', forthcoming Special Issue *Climate Change Economics*.

Eom, J., J. Edmonds, V. Krey, N. Johnson, T. Longden, G. Luderer, K. Riahi and D.P. Van Vuuren (2014), 'The impact of near-term climate policy choices on technology and emission transition pathways', forthcoming Special Issue *Technological Forecasting and Social Change*.

Favero, A. and E. Massetti (2014), 'Trade of woody biomass for electricity generation under climate mitigation policy', forthcoming *Resource and Energy Economics*, Supersedes Nota di Lavoro 13.2013, Milan: Fondazione Eni Enrico Mattei.

Kriegler, E., K. Riahi, N. Bauer, V.J. Schwanitz, N. Petermann, V. Bosetti, A. Marcucci, S. Otto, L. Paroussos, S. Rao, T. Arroyo Curras, S. Ashina, J. Bollen, J. Eom, M. Hamdi-Cherif, T. Longden, A. Kitous, A. Méjean, F. Sano, M. Schaeffer, K. Wada, P. Capros, D.P. Van Vuuren and O. Edenhofer (2014), 'Making or breaking climate targets: The AMPERE study on staged accession scenarios for climate policy', forthcoming Special Issue *Technological Forecasting and Social Change*.

Kriegler, E., M. Tavoni, T. Aboumahboub, G. Luderer, K. Calvin, G. De Maere, V. Krey, K. Riahi, H. Rosler, M. Schaeffer and D. Van Vuuren (2014), 'What does the 2°C target imply for a global climate agreement in 2020? The LIMITS study on Durban Platform scenarios', forthcoming Special Issue *Climate Change Economics*.

McCollum, D., Y. Nagai, K. Riahi, G. Marangoni, K. Calvin, R. Pietzcker, J. Van Vliet and B. Van Der Zwaan (2014), 'Energy investments under climate policy: a comparison of global models', forthcoming Special Issue *Climate Change Economics*.

Marangoni, G. and M. Tavoni (2014), 'The clean energy R&D strategy for 2°C', forthcoming Special Issue *Climate Change Economics*.

Pietzcker, R., T. Longden, W. Chen, S. Fu, E. Kriegler, P. Kyle and G. Luderer (2014), 'Long-term transport energy demand and climate policy: Alternative visions on transport decarbonization in energy-economy models', forthcoming *Energy*. Supersedes Nota di Lavoro 8.2013, Milan: Fondazione Eni Enrico Mattei.

Riahi, K., E. Kriegler, N. Johnson, C. Bertram, M. Den Elzen, J. Eom, M. Schaeffer, J. Edmonds, M. Isaac, V. Krey, T. Longden, G. Luderer, A. Méjean, D.L. McCollum, S. Mima, H. Turton, D.P. Van Vuuren, K. Wada, V. Bosetti, P. Capros, P. Criqui and M. Kainuma (2014), 'Locked into Copenhagen pledges: Implications of short-term emission targets for the cost and feasibility of long-term climate goals', forthcoming *Technological Forecasting and Social Change*.

Tavoni, M, E. Kriegler, T. Aboumahboub, K. Calvin, G. De Maere, J. Jewell, T. Kober, P. Lucas, G. Luderer, D. McCollum, G. Marangoni, K. Riahi and D. Van Vuuren (2014), 'The distribution of the major economies effort in the Durban platform scenarios', forthcoming Special Issue *Climate Change Economics*.

Van Der Zwaan, B., H. Rösler, T. Kober, T. Aboumahboub, K. Calvin, D. Gernaat, G. Marangoni and D. McCollum (2014), 'A cross-model comparison of global long-term technology diffusion under a 2°C climate change control target', forthcoming Special Issue *Climate Change Economics*.

2013

Bosello, F., C. Carraro and E. De Cian (2013), 'Adaptation can help mitigation: An integrated approach to post-2012 climate policy', *Environment and Development Economics*, **18**(3), 270–290, dx.doi.org/10.1017/S1355770X13000132. Supersedes Nota di Lavoro 69.2011, Milan: Fondazione Eni Enrico Mattei.

Bosetti, V. and E. De Cian (2013), 'A good opening: the key to make the most of unilateral climate action', *Environmental and Resource Economics*, **56**(2), 255–276, dx.doi.org/10.1007/s10640-013-9643-1. Supersedes Nota di Lavoro 81.2011, Milan: Fondazione Eni Enrico Mattei.

Bosetti, V. and T. Longden (2013), 'Light duty vehicle transportation and global climate policy: The importance of electric drive vehicles', *Energy Policy*, **58**, 209–221. Supersedes Nota di Lavoro 11.2012, Milan: Fondazione Eni Enrico Mattei.

Bosetti, V., C. Carraro, E. De Cian, E. Massetti and M. Tavoni (2013), 'Incentives and stability of international climate coalitions: An integrated assessment', *Energy Policy*, **55**, 44–56, dx.doi.org/10.1016/j.enpol.2012.12.035. Supersedes Nota di Lavoro 97.2011, Milan: Fondazione Eni Enrico Mattei.

Calvin, K., S. Pachauri, E. De Cian and I. Mouratiadou (2013), 'The effect of African growth on future global energy, emissions, and regional development', forthcoming Special Issue *Climatic Change*, dx.doi.org/10.1007/s10584-013-0964-4.

Carrara, S. and G. Marangoni (2013), 'Non-CO_2 greenhouse gas mitigation modelling with marginal abatement cost curves: technical change, emission scenarios and policy costs', *Economics and Policy of Energy and the Environment*, **55**(1), 91–124, dx.doi.org/10.3280/EFE2013-001006.

Carraro, C., E. De Cian and M. Tavoni (2013), 'Human capital, innovation, and climate policy: An integrated assessment', *Environmental Modeling and Assessment*, 10.1007/s10666-013-9385-z. Supersedes Nota di Lavoro 18.2012, Milan: Fondazione Eni Enrico Mattei.

Cherp, A., J. Jewell, V. Vinichenko, N. Bauer and E. De Cian (2013), 'Global energy security in long-term scenarios under different climate policies, GDP growth and fossil fuel availability assumptions', forthcoming Special Issue *Climatic Change*, dx.doi.org/10.1007/s10584-013-0950-x.

De Cian, E., I. Keppo, S. Carrara, K. Schumacher, H. Förster, M. Hübler, J. Bollen and S. Paltsev (2013), 'European-led climate policy versus global mitigation action. Implications on trade, technology, and energy', forthcoming Special Issue *Climate Change Economics*, **4**(4)

De Cian, E., F. Sferra and M. Tavoni (2013), 'The influence of economic growth, population, and fossil fuel scarcity on energy investments', forthcoming special Issue *Climatic Change*, dx.doi.org/10.1007/s10584-013-0902-5. Supersedes Nota di Lavoro 59.2013, Milan: Fondazione Eni Enrico Mattei.

De Cian, E., S. Carrara and M. Tavoni (2013), 'Innovation benefits from nuclear phase-out: can they compensate the costs?', forthcoming Special Issue *Climatic Change*, dx.doi.org/10.1007/s10584-013-0870-9. Supersedes Nota di Lavoro 96.2012, Milan: Fondazione Eni Enrico Mattei.

Förster, H., K. Schumacher, E. De Cian, M. Hübler, I. Keppo, S. Mima and R.D. Sands (2013), 'European energy efficiency and decarbonization strategies beyond 2030 – A sectoral multi-model decomposition', forthcoming Special Issue *Climate Change Economics*, **4**(4).

Knopf, B., H. Chen, E. De Cian, H. Förster, A. Kanudia, I. Karkatsouli, I. Keppo, T. Koljonen, K. Schumacher, D.P. Van Vuuren (2013), 'Beyond 2020 – European strategies and costs for an energy system transformation', forthcoming Special Issue *Climate Change Economics*, **4**(4).

Kriegler, E., N. Petermann, V. Krey, V.J. Schwanitz, G. Luderer, S. Ashina, V. Bosetti, J. Eom, A. Kitous, A. Mejean, L. Paroussos, F. Sano, H. Turton, C. Wilson and D.P. Van Vuuren (2013), 'Diagnostic indicators for integrated assessment models of climate policies', forthcoming Special Issue *Technological Forecasting and Social Change*.

Luderer, G., C. Bertram, K. Calvin, E. De Cian and E. Kriegler (2013), 'Implications of weak near-term climate policies on long-term climate mitigation pathways', forthcoming Special Issue *Climatic Change*. dx.doi.org/10.1007/s10584-013-0899-9.

Massetti, E. and E.C. Ricci (2013), 'An assessment of the optimal size and timing of investments in concentrated solar power', *Energy Economics*, **38**, 186–203, dx.doi.org/10.1016/j.eneco.2013.02.012. Supersedes Nota di Lavoro 47.2011, Milan: Fondazione Eni Enrico Mattei.

2012

Bosetti, V., C. Carraro and M. Tavoni (2012), 'Timing of mitigation and technology availability in achieving a low-carbon world', *Environment and Resource Economics*, **51**(3), 353–369, dx.doi.org/10.1007/s10640-011-9502-x.

Bosetti, V. and J. Frankel (2012), 'Politically feasible emission target formulas to attain 460 ppm CO_2 concentrations', *Review of Environmental Economics and Policy*, **6**(1), 86–109, dx.doi.org/10.1093/reep/rer022. Supersedes Nota di Lavoro 92.2009, Milan: Fondazione Eni Enrico Mattei.

Carraro, C. and E. Massetti (2012), 'Energy and climate change in China', *Environment and Development Economics*, 17(6), 689–713, dx.doi.org/10.1017/S1355770X12000228. Supersedes Nota di Lavoro 16.2011, Milan: Fondazione Eni Enrico Mattei.

Carraro, C., A. Favero and E. Massetti (2012), 'Investments and public finance in a green, low carbon, economy', *Energy Economics*, **34**(Supplement 1), S15–S28. dx.doi.org/10.1016/j.eneco.2012.08.036.

Carraro, C. and E. Massetti (2012), 'Editorial', Special Issue *International Environmental Agreements, Law, Economics and Politics*, **11**(3), 205–208. dx.doi.org/10.1007/s10784-011-9160-z.

Carraro, C. and E. Massetti (2012), 'Beyond Copenhagen: A realistic climate policy in a fragmented world', *Climatic Change*, **110**(3), 523–542, dx.doi.org/10.1007/s10584-011-0125-6. Supersedes Note di Lavoro 136.2010, Milan: Fondazione Eni Enrico Mattei.

De Cian, E. and M. Tavoni (2012), 'Mitigation portfolio and policy instruments when hedging against climate policy and technology uncertainty', *Environmental Modeling & Assessment*, **17**(1-2), 123–136,. dx.doi.org/10.1007/s10666-011-9279-x.

De Cian, E. and M. Tavoni (2012), 'Do technology externalities justify restrictions on emission permit trading?', *Resource and Energy Economics*, **34**(4), 624–646, dx.doi.org/10.1016/j.reseneeco.2012.05.009. Supersedes Nota di Lavoro 33.2011, Milan: Fondazione Eni Enrico Mattei.

Höhne, N., C. Taylor, R. Elias, M. Den Elzen, K. Riahi, C. Chen, J. Rogelj, G. Grassi, F. Wagner, K. Levin, E. Massetti and Z. Xiusheng (2012), 'National greenhouse gas emissions reduction pledges and 2°C – Comparison of studies', *Climate Policy*, **12**(3), 356–377, dx.doi.org/10.1080/14693062.2011.637818.

Massetti, E. (2012), 'Short-term and long-term climate mitigation policy in Italy', *WIREs Climate Change*, **3**(2), 171–183. dx.doi.org/10.1002/wcc.159.

Massetti, E. and M. Tavoni (2012), 'A developing Asia emission trading scheme (Asia ETS)', *Energy Economics*, **34**(Supplement 3), S436–S443. dx.doi.org/10.1016/j.eneco.2012.02.005.

Massetti, E. (2012), 'Carbon tax scenarios for China and India: Exploring politically feasible mitigation goals', Special Issue *International Environmental Agreements, Law, Economics and Politics*, **11**(3), 209–227, dx.doi.org/10.1007/s10784-011-9157-7. Supersedes Nota di Lavoro 24.2011, Milan: Fondazione Eni Enrico Mattei.

2011

Bosetti, V., C. Carraro, R. Duval and M. Tavoni (2011), 'What should we expect from innovation? A model-based assessment of the environmental and mitigation cost implications of climate-related R&D', *Energy Economics*, **33**(6), 1313–1320, dx.doi.org/10.1016/j.eneco.2011.02.010. Supersedes Nota di Lavoro 42.2010, Milan: Fondazione Eni Enrico Mattei.

Bosetti, V., R. Lubowski, A. Golub and A. Markandya, (2011), 'Linking reduced deforestation and a global carbon market: implications for clean energy technology and policy flexibility', *Environment and Development Economics*, **16**(4), 479–505, dx.doi.org/10.1017/S1355770X10000549. Supersedes Nota di Lavoro 56.2009, Milan: Fondazione Eni Enrico Mattei.

Bosetti, V. (2011), 'Which technologies should we pick? A comment on "Technology Interactions for Low Carbon Energy Technologies: What can we learn from a larger number of scenarios" by Haewon Chon, Leon Clarke, Page Kyle, Marshall Wise, Andy Hackbarth and Robert Lempert', *Energy Economics*, **33**(4), 632–633. dx.doi.org/10.1016/j.eneco.2010.11.008.

Bosetti, V. and D. Victor (2011), 'Politics and economics of second-best regulation of greenhouse gases: The importance of regulatory credibility', *The Energy Journal*, **32**(1), 1–24. Supersedes Nota di Lavoro 29.2010, Milan: Fondazione Eni Enrico Mattei.

Massetti, E. and M. Tavoni (2011), 'The cost of climate change mitigation policy in Eastern Europe, Caucasus and Central Asia', *Climate Change Economics*, **2**(4), 341–370, dx.doi.org/10.1142/S2010007811000346.

Tavoni, M. and B. Van Der Zwaan (2011), 'Nuclear versus coal plus CCS: A comparison of two competitive base-load climate control options', *Environmental Modeling and Assessment*, **16**(5), 431–440, dx.doi.org/10.1007/s10666-011-9259-1. Supersedes Nota di Lavoro 100.2009, Milan: Fondazione Eni Enrico Mattei.

2010

Bosello, F., C. Carraro and E. De Cian (2010), 'Climate policy and the optimal balance between mitigation, adaptation and unavoided damage', *Climate Change Economics*, **1**(2), 71–92, dx.doi.org/10.1142/s201000781000008x. Supersedes Nota di Lavoro 32.2010, Milan: Fondazione Eni Enrico Mattei.

2009

Bosetti, V., C. Carraro and M.Tavoni (2009), 'A Chinese commitment to commit: Can it break the negotiation stall?', *Climatic Change*, **97**(1–2), 297–303, dx.doi.org/10.1007/s10584-009-9726-8. Supersedes Nota di Lavoro 32.2010, Milan: Fondazione Eni Enrico Mattei.

Bosetti, V. and B. Van Der Zwaan (2009), 'Targets and technologies for climate control', *Climatic Change*, **96**(3), 269–273, dx.doi.org/10.1007/s10584-009-9631-1.

Bosetti, V., C. Carraro, A. Sgobbi and M. Tavoni (2009), 'Delayed action and uncertain stabilisation targets. How much will the delay cost?', *Climatic Change*, **96**(3), 299–312, dx.doi.org/10.1007/s10584-009-9630-2. Supersedes Nota di Lavoro 69.2008, Milan: Fondazione Eni Enrico Mattei.

Bosetti, V. and B. Buchner (2009), 'Data envelopment analysis of different climate policy scenarios', *Ecological Economics*, **68**(5), 1340–1354, dx.doi.org/10.1016/j.ecolecon.2008.09.007. Supersedes Nota di Lavoro 82.2005, Milan: Fondazione Eni Enrico Mattei.

Bosetti, V., C. Carraro and M.Tavoni (2009), 'Climate change mitigation strategies in fast-growing countries: The benefits of early action', *Energy Economics*, **31**(S2), S144–S151, dx.doi.org/10.1016/j.eneco.2009.06.011. Supersedes Nota di Lavoro 53.2009, Milan: Fondazione Eni Enrico Mattei.

Bosetti, V., C. Carraro, E. Massetti, A. Sgobbi and M. Tavoni (2009), 'Optimal energy investment and R&D strategies to stabilise greenhouse gas atmospheric concentrations', *Resource and Energy Economics*, **31**(2), 123–137, dx.doi.org/10.1016/j.reseneeco.2009.01.001. Supersedes Nota di Lavoro 95.2007, Milan: Fondazione Eni Enrico Mattei.

Bosetti, V., C. Carraro and M.Tavoni (2009), 'Climate policy after 2012. Technology, timing, participation', *CESifo Economic Studies*, **55**(2), 235–254, dx.doi.org/10.1093/cesifo/ifp007.

Bosetti, V., C. Carraro and E. Massetti (2009), 'Banking permits: Economic efficiency and distributional effects', *Journal of Policy Modeling*, **31**(3), 382–403, dx.doi.org/10.1016/j.jpolmod.2008.12.005. Supersedes Nota di Lavoro 01.2008, Milan: Fondazione Eni Enrico Mattei.

Bosetti, V. and M. Tavoni (2009), 'Uncertain R&D, backstop technology and GHGs stabilization', *Energy Economics*, **31**(S1), S18–S26, dx.doi.org/10.1016/j.eneco.2008.03.002. Supersedes Nota di Lavoro 6.2007, Milan: Fondazione Eni Enrico Mattei.

Carraro, C., E. Massetti and L. Nicita (2009), 'How does climate policy affect technical change? An analysis of the direction and pace of technical progress in a climate-economy model', *The Energy Journal*, **30**(SI2), 7–37, dx.doi.org/10.5547/ISSN0195-6574-EJ-Vol30-NoSI2-2. Supersedes Nota di Lavoro 8.2009, Milan: Fondazione Eni Enrico Mattei.

2008

Bosetti, V., C. Carraro, E. Massetti and M. Tavoni (2008), 'International energy R&D spillovers and the economics of greenhouse gas atmospheric stabilization', *Energy Economics*, **30**(6), 2912–2929, dx.doi.org/10.1016/j.eneco.2008.04.008. Supersedes Nota di Lavoro 82.2007, Milan: Fondazione Eni Enrico Mattei.

2007

Tavoni, M., B. Sohngen and V. Bosetti (2007), 'Forestry and the carbon market response to stabilize climate', *Energy Policy*, **35**(11), 5346–5353, dx.doi.org/10.1016/j.enpol.2006.01.036. Supersedes Nota di Lavoro 15.2007, Milan: Fondazione Eni Enrico Mattei.

2006

Bosetti, V., C. Carraro, M. Galeotti, E. Massetti and M. Tavoni (2006), 'WITCH: A World Induced Technical Change Hybrid Model', *The Energy Journal*, special issue on 'Hybrid modeling of energy- environment policies: Reconciling bottom-up and top-down', **27**(SI2), 13–38.

Index